Development Geography

The poor, equity, social justice and the environment are issues which continue to motivate large numbers. The rapidly increasing globalisation of economic, social and political life now demands that we gain a wider and deeper understanding of and perspective on the issues that face us all.

Development Geography has been written to stimulate critical thought and discussion about development. It does not assume any clear-cut distinction between 'developed' and 'developing' parts of the world. There are pockets of poverty and low levels of development in the richest countries, just as there are pockets of wealth and high levels of development in the poorest countries. Similarities between developed and developing countries extend to embrace institutions, values, patterns of behaviour and concepts of prestige, wealth and power. The developed and developing worlds are not as distant from each other as some would like to think.

Development Geography is an informative and vibrant introductory text, with a wealth of contrasting case studies and illustrations. It facilitates a more integrated understanding of development and places development problems within the context of the global economy and society. Topics covered range from population and culture to agricultural and industrial development. Case studies highlight the possible solutions, as well as the problems, at local, national and international levels.

Rupert Hodder is Senior Lecturer in Geography at the University of Plymouth.

Routledge Contemporary Human Geography Series

Series Editors:
David Bell and **Stephen Wynn Williams**, Staffordshire University

This series of texts offers stimulating introductions to the core subdisciplines of human geography. Building between 'traditional' approaches to subdisciplinary studies and contemporary treatments of these same issues, these concise introductions respond particularly to the new demands of modular courses. Uniformly designed, with a focus on student-friendly features, these books form a coherent series which is up-to-date and reliable.

Existing titles:

Urban Geography

Cultural Geography

Tourism Geography

Techniques in Human Geography

**Routledge Contemporary Human
Geography Series**

Development Geography

Rupert Hodder

London and New York

First published 2000
by Routledge
2 Park Square, Milton Park, Abingdon, Oxon, OX14 4RN

Simultaneously published in the USA and Canada
by Routledge
270 Madison Ave, New York NY 10016

Routledge is an imprint of the Taylor & Francis Group

Transferred to Digital Printing 2007

Typeset in Times by Keystroke, Jacaranda Lodge, Wolverhampton

British Library Cataloging in Publication Data
A catalogue record for this book is available from the British Library

Library of Congress Cataloging in Publication Data
Hodder, Rupert.
     Development geography / Rupert Hodder.
        p. cm. — (Routledge contemporary human geography series)
     Includes bibliographical references and index.
     1. Economic development. 2. Economic geography.
     I. Title. II. Series.
HD75.H618 2000
338.9—dc21   00–021338

ISBN 0–415–14210–5 (hbk)
ISBN 0–415–14211–3 (pbk)

Printed and bound by CPI Antony Rowe, Eastbourne

Contents

Plates

Figures

Tables

Boxes

Acknowledgements

My thanks to Mark Blacksell, Mark Cleary, Richard Gibb, David Pinder and others in my Department for their help and support; and to my students at Plymouth, the Chinese University of Hong Kong and the London School of Economics for their criticisms, views and arguments. I would also like to thank Brian Rogers and Tim Absalom for Figure 1.1.

The author and publisher would like to thank the following for permission to reproduce copyright material: Panos Pictures for Plates 2, 3, 4, 7, 13, 14, 15, 16; B.W. Hodder for Plates 1, 5, 6, 8, 9, 10; R.N.W. Hodder for Plates 11 and 12; the Editor of *Third World Planning Review* for permission to use Figure 5.2.

Abbreviations

ADB	Asian Development Bank
AIDS	acquired immunodeficiency syndrome
APEC	Asia-Pacific Economic Cooperation Forum
ASEAN	Association of South East Asian Nations
CBD	central business district
CIS	Commonwealth of Independent States
DPEP	District Primary Education Project (South Africa)
ECOWAS	Economic Community of West African States
EIO	export-oriented industrialisation
EPZ	export processing zone
EU	European Union
FAO	United Nations Food and Agricultural Organization
FDI	foreign direct investment
G7	Group of Seven Leading Industrial Nations
GATT	General Agreement on Tariffs and Trade
GDI	Gender-related Development Index
GDP	gross domestic product
GEM	Gender Empowerment Measure
HDI	Human Development Index
HIPC	heavily indebted poor country
HIV	human immunodeficiency virus
HYV	higher-yielding variety (of a crop)
IMF	International Monetary Fund
ISI	import substitution industrialisation
ITDG	Intermediate Technology Development Group
LAFTA	Latin America Free Trade Association
MNC	multinational corporation
NAFTA	North America Free Trade Association
NGO	non-governmental organisation
NIC	newly industrialising country (*or* newly industrialised country)

NIDL	new international division of labour
ODA	Overseas Development Administration (UK)
OECD	Organisation for Economic Co-operation and Development
OPEC	Organisation of Petroleum Exporting Countries
PPP	purchasing power parity
SADCC	Southern African Development Coordination Conference
SAP	structural adjustment programme
SEA	strategic environmental assessment
SEZ	Special Economic Zone (China)
SOE	state-owned enterprise
TNC	transnational corporation
UNCED	United Nations Conference on Environment and Development
UNDP	United Nations Development Programme
UNHCR	United Nations High Commissioner for Refugees
UNICEF	United Nations Children's Fund
WTO	World Trade Organization

Approaches to development

- Outline of approach
- Definitions of 'development' and 'the developing world'
- Theories of development

Outline of approach

This short book is designed to stimulate critical thought and discussion about development and, in particular, about development in what is conventionally termed the developing world. The need for such thought and discussion has never been greater. Travelling and learning about faraway and exotic places has long been an attractive occupation, and more and more people are now able to travel and stay in the world's poorest countries. Concern for the poor, equity and social justice and for the environment in which the poor of the world live has motivated and continues to motivate large numbers of people. But the rapidly increasing globalisation of economic, social and political life now demands a wider and deeper understanding of and perspective on the issues that face the world's poor. There is a need for everyone to challenge many of the views, ideas and opinions which too often constitute the received wisdom or thinking about the developing world today. Moreover, the former Soviet Union, together with much of Eastern Europe, and China, are now usually counted as developing economies, and their central role in international affairs and security demands much greater knowledge and understanding in advanced countries, more particularly in the most powerful economies of the United States, Western Europe and Japan.

Three further points need to made clear about the approach adopted in this book. First, the book is necessarily very selective in its coverage, not attempting to provide a comprehensive geography of the developing world. A limit on length makes this impossible, and in any case such a task lies outside the aims and purpose of the Routledge Contemporary Human Geography series, of which this book is a part. Second, this text does not assume any clear-cut distinction between the 'developed' and 'developing' parts of the world. Some would go so far as to argue that the distinction no longer has any meaning, and, indeed, that it perpetuates the damaging 'rich and poor' divide in our thinking and confirms a certain patronising attitude towards the world's poor. This viewpoint will be raised again at the end of the book. For the present, and as a study of the distribution and explanation of relative poverty and global inequalities in levels of development, the emphasis in these pages is inevitably on those parts of the world where poverty is most widespread and evident. To some degree, however, most, if not all, aspects of development geography apply universally: there are pockets of poverty and low levels of development in the richest countries, just as there are pockets of wealth and high levels of development in the poorest countries. Similarities between developed and developing countries also extend beyond the existence of islands of poor among the rich and of rich among the poor to embrace institutions, values, attitudes, patterns of behaviour and concepts of prestige, wealth and power. Reference is made to these similarities throughout this book and should help to avoid the common danger of a reader 'switching off' all experience and knowledge of the developed world while thinking about the developing world. It also facilitates the placing of development problems within the context of the global economy and society, and enables reference to be made, for example, to development problems in the former Soviet Union and Eastern Europe, and to the role of North America, Western Europe and Japan in reducing (or perpetuating) poverty in the developing world.

Third, the reader is encouraged to challenge many of the assumptions about development and, in particular, about the causes of poverty. The media continue to play an important role in informing us about the problems of the developing world and in encouraging a deep and often active concern for the plight of so many people in the poorest countries. Nevertheless, there is some danger that information and opinions can be too easily accepted uncritically as 'facts', or at least as received wisdom. This applies to development theory as much as to practical policies. No attempt is made here to direct the reader along any particular theoretical or ideological path. While some of the main theoretical approaches currently under discussion in the literature are introduced in this chapter and referred to where appropriate throughout the book, it is hoped that readers will reach their independent conclusions about the validity of the various elements in the constantly shifting theoretical debate in development studies.

Before tackling the problems of definition, it is important to make the point that there can be no 'correct' definition of 'development' or 'the developing world'. The perspectives from which to approach development geography are many, varied and often conflicting, so that almost all statements made in these pages can be challenged by someone whose perspectives, assumptions and ideology are different from my own. Development studies generally is a fertile field for radical debate of one form or another; and geographers, like all scholars, 'are prone to sinking their feet into the quagmire of definitions' (Rigg *et al.* 1999: 581). As one group of authors puts it,

> around the university world, in departments of anthropology, economics, development studies, geography and political economy, first-year students are being asked to think, debate and write about the meaning of 'development'. As they sit earnestly at their first lectures . . . they are told that development is one of the hardest of terms to define.
>
> (ibid.: 583–4)

Nevertheless, it is important to have some working definitions of 'development' and 'the developing world' in mind if any kind of clarity is to be achieved. All one can do in approaching the study of development geography is to start from what appears to be a sensible working definition while being prepared to reconsider one's position as circumstances, evidence or experience suggest.

Definitions

'Development'

There is some, though by no means universal, agreement that the term 'development' refers to a process which results in economic growth. But this is a very narrow definition based solely on economic criteria, and it is usual to include other, non-economic criteria and concepts. Thus development can be defined as an economic, social and political process which results in a cumulative rise in the perceived standard of living for an increasing proportion of a population. Such a definition suggests that an increased standard of living involves a social and political process as well as an economic one; that a rise in the standard of living must be cumulative, rather than just temporary; and that it has to be significant enough to be perceived by increasing numbers in a population. Clearly, this kind of definition embraces the concepts of equality and equity as crucial aims in any development process, but it does not go so far as to place the removal of all inequalities as the central aim in the development process. More simply, some argue that development is best defined as 'growth with equity'.

Other definitions focus on technological change as a significant criterion. Thus development can be defined as the process by which a traditional society employing traditional, unsophisticated techniques is transformed into a modern, high-technology, high-income economy in which capital, labour skills and scientific knowledge replace labour-intensive methods of production. Clearly, the difficulties of trying to embody all possible criteria in a simple definition of development are immense, and one way of avoiding this problem is simply to describe the characteristics of less developed as distinct from developed economies. Thus Todaro (1997: 38) suggests that less developed economies are characterised by low levels of living, reflected in such indices as low income, high inequalities, poor health and inadequate education; by low levels of productivity; by high rates of population growth and dependency burdens; by high and rising levels of unemployment and underemployment; by substantial dependence on agricultural production and primary export products; by the prevalence of imperfect markets and limited information; and by dominance, dependence and vulnerability in international relations. However, it is easy enough to find exceptions, even to this broad list of characteristics; and though this descriptive definition has its uses in providing a checklist, it does not bring us nearer to a wholly satisfactory definition of development. Other writers place emphasis on the fact that many developing countries were formerly colonial territories, with all colonialism's predictable economic, social, political and psychological effects. Development may also have a political perspective: that development should encompass freedom of action and expression, enabling people to have wider choices (Dickenson *et al.* 1996: 28). Or emphasis may be laid on three major objectives of development: sustenance, self-esteem and the ability to choose (Todaro 1997: 18).

Perhaps it is best to conclude that a wholly acceptable and universally applicable theory of development is neither possible nor desirable. As one writer has put it,

development is historical, diverse, complex and contradictory; it is the
central feature of the human condition. To reduce it to a number of
asocial characteristics and their interaction is to trivialize the experience
of real societies and the struggle of their peoples to make a living.

(RL in Johnston *et al.* 1996: 130)

It is also important to realise that development can be approached from
many different directions. Its study is a vast, interdisciplinary field and it
attracts and demands the participation of workers whose interests and
training are focused in or originate from many different disciplines. In
any study of development it is sometimes difficult to decide which
discipline is the most appropriate or relevant: economics, sociology,
agricultural science, politics, geography or any other field of study. In
practice, however, this does not raise any serious difficulties. In the case
of geography, for instance, the geographer's special interest in space,
place and the environment directs, or at least informs, the issues and
problems he feels most comfortable in examining. Furthermore, the
geographer, like practitioners from all other disciplines, is always alive to
the fact that in the study of development his or her work should be
regarded as complementary to the work of colleagues from other
disciplines.

'The developing world'

'The developing world' is the term used here to refer to what is otherwise
known as the 'third', 'less developed'; 'developing', 'undeveloped',
'underdeveloped' or 'backward' world; other terms include 'emerging
economies', 'transitional economies' and 'the South'. Some of these
terms have slightly different connotations, but for practical purposes they
may be regarded as synonymous. All are in some ways unsatisfactory,
even misleading, and it is becoming increasingly clear that we need to
change quite fundamentally our thinking about the geographical division
of the world implied in, for instance, the term 'Third World'. Thus it has
been pointed out that the division of the world into 'First', 'Second' and
'Third' worlds is now outmoded, because the past two decades

have seen the dramatic emergence of the newly industrialized economies
[NICs] of South East Asia, where growth rates have far outstripped those
of the advanced countries. . . . Equally profound has been the collapse of
many of the former socialist countries, and their difficult transition to

capitalism. In both cases, these new spaces of capitalism have proved
problematic . . . crises in one area of the world now quickly spill over into
other areas far removed from the original sources of instability. Yet, these
instabilities notwithstanding, there can be no doubt that we are entering a
new age in which the trajectory of the global economy will be much more
centred on the development of Asian capitalism (especially China and
India). Not only are there various shifts in the global spread of capitalism
fostering all sorts of regional-bloc trading and integration agreements
across the developed and developing world, their impacts at the local
level are likely to be profound.

(Bryson *et al.* 1999: 13)

In these pages use is made of the classification of world economies
adopted by the World Bank in its annual *World Development Report*.
Figure 1.1 gives the latest picture from this source (World Bank 1998a).
The data on which this figure is based have severe limitations, relying as
they do on one criterion – per capita income – but they are simple and
useful as long as they are taken in conjunction with other indices of
development available in the *World Development Report*. This
classification identifies four groups of economies (Table 1.1). The
poorest of these – 'low-income economies' (such as Bangladesh or
Nicaragua) – are defined as having GNP per capita incomes of less than
$785; those economies with per capita incomes above this figure but
below $3 125 are classified as 'low-middle-income' economies
(including Egypt or Costa Rica); while those economies with per capita
incomes above that figure but below $9 655 are classified as
'upper-middle-income' economies (such as Mexico or Brazil). Finally,
those economies with per capita incomes above $9 655 are termed
'high-income economies' (for example, the United States or the United
Kingdom). In a general sense, the developing world embraces the first
three of these four groups. But, as Table 1.1 shows, there is some overlap
between the third and fourth groups. It is clear that the main
concentration of economies is at the lowest (poorest) end of the range
and that they include a disproportionate number of African economies
and disproportionately few Latin American economies.

These countrywide figures ignore the great range of levels of wealth and
poverty within economies: spatial inequalities at different scales, urban
and rural, by regions, classes and gender. This classification also depends
on average per capita figures as the sole criterion. On this basis China
comes well down the spectrum but, taking local purchasing power of
currencies into account, China's economy looks much bigger. According

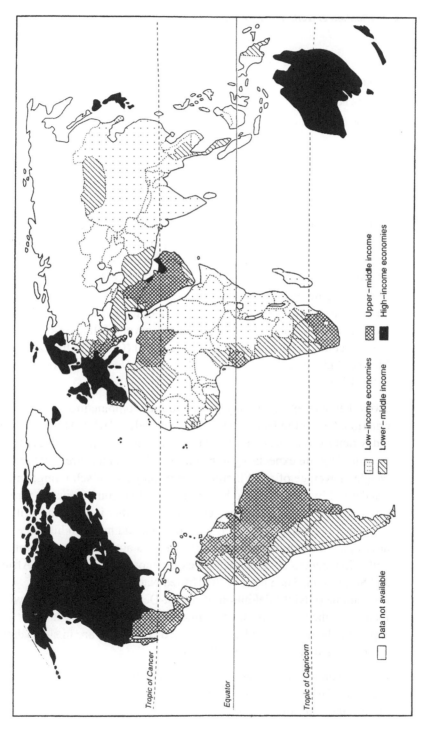

Figure 1.1 *The four groups of world economies, by per capita income*

Source: After World Bank (1999)

Table 1.1 *Classification of world economies into four groups, 1997*

	Population (millions)	GNP ($ billion)	GNP ($ per capita)
Low-income countries	2 048	722	350
Lower middle-income countries	2 285	2 818	1 230
Upper middle-income countries	571	2 584	4 520
High-income countries	926	23 802	25 700
World (total)	5 829	29 926	5 130

Source: World Bank (1999)

to the World Bank's latest World Development Indicators, China's GDP in 1997, measured at purchasing-power parity (PPP), was $3.8 trillion. This makes it the second largest economy in the world, ahead of Japan's $3.1 trillion, and roughly half of the United States' GDP of $7.8 trillion. Most emerging market economies look relatively larger when measured on a PPP basis, because their exchange rates tend to be undervalued relative to the dollar. Indeed, on this basis, seven of the world's fifteen largest economies are in the developing world (China, India, Brazil, Mexico, Indonesia, Russia and South Korea) (*The Economist*, 8 May 1999: 142).

Two other attempts at classifying economies combine more than one criterion. One is the Human Development Index (HDI). This generalised index ranks economies on a scale based on three criteria: longevity, measured by life expectancy at birth; knowledge, measured by the weighted average of adult literacy and mean years of schooling; and standard of living, measured by real per capita income adjusted for the differing PPP of each country's currency and other issues. The striking point about this classification is that the high and medium human development economies are in Latin America and that the economies with a low human development index are most characteristic of Africa and South Asia. Then there is the United Nations Development Programme (UNDP) definition of 'Least-Developed' economies, which focuses on the very poorest economies. In this case, as might be expected, the majority of least developed economies are in sub-Saharan Africa, with a few cases in South Asia.

Finally, mention should be made of the UN classification, which embraces the forty-four 'least-developed' economies as one of three groups, the other two being the eighty-eight non-oil-exporting developing countries and the thirteen petroleum-rich members of the Organization of Petroleum Exporting Countries (OPEC).

None of these classifications, attempting to give some precision to what is meant by the developing world, is entirely satisfactory. Which definition to choose must to some extent depend on the use to which any analysis is to be put. All suffer from incomplete, subjective or unreliable data. Not only are there great variations in the dates and reliability of statistical material, but some countries do not publish data regularly, or even at all. Thus in official statistical series, such as those of the World Bank, there are serious gaps in the information concerning China, Vietnam, Taiwan and Brunei. In some cases it is possible to fill in these gaps from local sources, but the problem of reliable comparative data remains. There are also difficulties arising from the different criteria used by different governments. In China, for example, GDP figures refer only to the gross value of industrial and agricultural output, ignoring services altogether.

On this basis, is 'the developing world' a valid or even a useful concept? Dickenson *et al.* (1996: 29–30) suggest that it is. They argue that international conditions, such as trade blocs, NICs and grassroots development, together with Friedmann's focus on the 'disempowered' (1966), emphasising people, not places, have all tended to affect our view of the developing world. Todaro supports this notion from the viewpoint of his own discipline, arguing that development economics is 'a field of study that is rapidly evolving its own distinctive analytic and methodological identity . . . it is the economics of contemporary poor, undeveloped Third World nations with varying ideological orientations' (Todaro 1997: 7). On the other hand, the arguments *against* treating the developing world as being in some fundamental sense a discrete, distinctive part of the world are strong. Perhaps the best working position on this matter is to accept that

> it is fitting to consider the developing world as a whole as long as we recognise that the problems faced by such countries are different in *scale*, rather than in *kind*, from those that are faced by the richer nations of the world.
>
> (Potter and Lloyd-Evans 1998: 25)

A final point on definitions. Implicit in much of the discussion in this section and, indeed, throughout this book is that there is an agreed definition of 'poverty'. But is it possible to define poverty, except in the context of a particular place, time or culture? 'Absolute poverty' should be easier to define, and one author has attempted to define it as 'a specific minimum level of income needed to satisfy the basic physical needs of food, clothes and shelter in order to ensure continued survival' (Todaro 1997: 43). But does this definition of absolute poverty apply equally to London and Calcutta? Perhaps of more value is the concept of 'relative poverty', discussed at some length by several authors. This issue applies to Europe and the United States. In Europe, as far as the European Commission is concerned, regions qualifying as 'poor' must have a GDP per head less than 75 per cent of the EU average. This includes the whole of Greece, most of Spain and Portugal, the former East Germany, southern Italy (the Mezzogiorno), and parts of Austria, Finland and the United Kingdom. In the United Kingdom, apart from Northern Ireland and the Scottish Highlands and Islands, four other regions qualify for the next round of EU aid: Merseyside, South Yorkshire, West Wales and the Valleys, and Cornwall (*The Economist*, 27 February 1999: 32). In the United States, too, the US Census Bureau calculates that the number of Americans living below the poverty line in 1997–8 was 13.3 per cent of the population. But '41% of those "poor" households own their own homes, 70% own a car; 97% have colour televisions, and two-thirds enjoy air-conditioning'. As for food, most 'poor' children are growing up to be 'one inch taller and ten pounds heavier than the GIs who stormed the beaches of Normandy in World War II' (*The Economist*, 3 October 1998: 68).

Within the developing world itself there are marked differences in levels of poverty. If we compare the years 1985 and 2000, the number of people in poverty (as defined by the World Bank) has fallen most sharply in East Asia. Progress has been fair in South Asia, modest in Latin America and the Caribbean, and negligible in the Middle East and North Afria. In sub-Saharan Africa, however, slow economic growth and rapid population growth have meant a significant increase in levels of poverty. It is estimated that sub-Saharan Africa now accounts for over 30 per cent of the developing world's poor, as against 16 per cent in 1985.

Associated with the whole issue of poverty – its identification, measurement, causes and treatment – has arisen a wide range of ideas, such as the basic needs approach and the 'bottom-up' and 'top-down' paradigms, many of which link into the theoretical debate on development.

Theories of development

There are many reasons why it is important to know something about the various theories that have grown up around this whole subject of development. As will become clear in these pages, anyone attempting to discuss the problems of poor or rich countries starts, however unconsciously, from a number of assumptions and theoretical positions. Almost every statement made in these pages can be challenged on the grounds that it reflects a particular perspective, attitude or theoretical position. It is important, too, to be able to recognise that clear theoretical positions are implied in a number of reports and conferences, notably *Limits to Growth* (Meadows *et al.* 1972), the Brandt Report (1980), the Brundtland Report (1987), the United Nations Conference on Environment and Development (UNCED) in Rio de Janeiro in 1992, and in the recent rise in influence of, and criticisms of, the non-governmental organisations (NGOs) and globalisation.

Moreover, part of the aim of theory in our case is to try to explain patterns of inequality: why one country has developed rapidly and successfully while another remains rooted in poverty. It is clear that no single or simple explanatory factor such as population size or density, climate or natural resources will do. Thus no correlation linking climate and levels of development can be adduced from the evidence: while many poor countries lie in the humid tropics, many do not. Similarly with mineral resources: whereas some countries, such as Brunei and Venezuela, have certainly been helped in their development by the possession of oil resources, other countries, such as Nigeria, have not.

Some understanding of theory is also important in any discussion of the European impact on development in developing countries, most of which have at some time been subjected to European colonial rule. Indeed, some of the most important theoretical work has arisen out of varying interpretations of the colonial and post-colonial experience in developing countries. Certainly the effects of colonial control have been fundamental and dramatic. These have included the creation of clearly defined territories within definite geographical boundaries. Economic effects have included the disrupting of predominantly subsistence economies, the introduction and development of cash crops, the introduction of monetary exchange economies, and the establishment of internal networks of transport and communications. But perhaps the most significant economic legacy left by colonial powers to developing nations was that the economies of the various states were very much the creation of the colonial countries which formerly controlled them.

As for the social and political effects of European colonial control, many writers point to the effects on general health and education, and on the creation of indigenous professional and administrative nuclei. Europeans also established a general peace and a basic structure of law, order and administration within definite territorial limits. Furthermore, specific systems of government, including those modelled on Western systems of parliamentary democracy, were set up, and common languages were introduced.

But such comments can be and have been interpreted both positively and negatively; and it is on quite opposing assessments of the colonial period that much of the theoretical debate on development has been spawned. On the one hand, there is the opinion that Europeans fulfilled a 'civilising' mission in the developing world, giving its peoples a sound basis for future social, economic and political progress in the modern world. On the other hand, there is the point of view that most of the Third World's problems have their origin in the exploitative and destructive nature of European colonialism; that Europe created the dependency burden under which poor countries now labour.

Over the past fifty years or so, the literature on development has been influenced by a wide range of theories, many of them reflecting differing interpretations of the colonial impact as well as the political affiliations and bias of the various authors. The sequence has been predictable, moving from theories and models derived from capitalist analyses to a series of Marxist or neo-Marxist theories. More recently the theoretical debate has changed in content and impetus, but it is still possible to find writers who adhere firmly to earlier theoretical positions. At present some theoretical positions are more fashionable than others, but it is important that every student of development should reach his or her own working theoretical position, while being prepared to change an opinion in due course. What is undesirable is that one adopts uncritically the fashionable theoretical prescription of the moment, or the deeply and strongly held opinions of any particular author or lecturer. This is a field of continuous change, and all that is certain is that the fashionable theory of today might well be rejected tomorrow.

This is not the place to summarise all the major development theories and their respective variants, but mention should be made of four broad groups of development theories: early 'modernisation' theories, dependency theories, neo-classical theory and recent theories.

Early 'modernisation' theories

There is a broad range of early theories including the linear stages theory of Rostow (1960) and the structural change notions of such writers as Lewis (1955). Modernisation theories arose out of the European and US experience and the belief in managed capitalist economic and social development. They also 'arose from a new spirit of optimism and anti-communism after 1945 and from the determination of poor countries to secure economic progress' (Dickenson *et al.* 1997: 18). They are based on the notion that developed societies are distinguished by their economic, social, cultural and political modernity, which contrasts markedly with traditional values in developing societies. Economic development consists of social and cultural changes as well as improved production techniques.

Development, then, simply involves modernisation – mechanisation, rapid industrialisation and the transfer of the underemployed rural population to the productive urban-industrial sector. These modernisation theories held some sway until the end of the 1950s, and argued that the circular causation of poverty can be broken by industrialisation (Myrdal 1957) and that diffusion, spread effects and trickle-down effects (Hirschman 1958) will bring about development.

Dependency

Dependency theories cover a wide range of viewpoints, and can be traced back to Latin America with the work of Frank (1967) and Baran (1973). They were rooted in neo-Marxist political theory. Other names associated with dependency approaches – at their height during the 1970s – are Amin (1976) and Wallerstein (1979). Also under this heading can be placed the 'development of underdevelopment' school. As already noted, dependency theory has as its historical basis the fact that large parts of the developing world were for long part of the empires of European states – notably Britain, France, Portugal, Spain and the Netherlands. The 'development of underdevelopment' theories are based on the assumption that progress in the developing world has been, and continues to be, impeded by forces (international and/or domestic) bent on the ongoing exploitation of the developing world – the 'periphery', its peoples and its resources – by 'the core' – that is, the metropolitan areas of America and Europe (Friedmann 1966). The focus of dependency theory, then, is not on the process of development but on the roots of underdevelopment. In this kind of analysis, emphasis is laid on 'the apparently endless vision of the Western will to develop the world' – and to develop it in its own image (Corbridge 1995: 8).

However, the dependency and underdevelopment schools have been unable either to advance significantly our understanding of the complexities of the present predicament in the developing world or to trace accurately the dynamics of development processes in the 1980s and 1990s. While popular with academics, this group of theories has had remarkably little influence on actual development strategies in less developed countries. On practical grounds those few countries, like Benin and several other African countries, which have adopted some form of neo-Marxist thinking in their economic strategies have tended to perform badly, while those countries, such as the East Asian NICs, which have rejected neo-Marxist approaches have done rather well. In East Asia there has been relatively little intense ideological critique of colonialism of the kind found in Latin America and Africa (Hodder 1992: 33). Other criticisms of the dependency approach include the opinion that 'the dependency position is vitiated by a variable combination of circular reasoning, fallacious inferences from empirical observations and a weak base in deductive theory' (Booth 1985: 762). Other criticisms of neo-Marxist views include those of Warren (1980), who accepts that post-imperialist capitalism promotes development in the developing world: 'capitalism is dynamic and developed in terms of economic growth, even if it is irrational and unfair in the manner in which it distributes the fruits of this growth' (Corbridge 1995: 6).

Neo-classical theory

Neo-classical theory has had a great influence on development thinking in the 1980s and 1990s. It has also influenced the World Bank and the International Monetary Fund (IMF) in their strategic adjustment programmes (SAPs); the IMF insists on certain policies being followed by recipient countries, and these conditions very much reflect the thinking of neo-classical counter-revolutionists. This school challenged

Plate 1 *European colonial church in Malacca, 1753*
Photograph: B.W. Hodder

earlier models: the state is viewed as a primary motor force behind social and economic occurrences, and state leaders are held responsible for the political and economic fortunes of a country. Unlike its predecessors, the modernisation and dependency schools, and in ongoing contrast to materialist studies, this school has broken out of existing moulds and placed political factors at the centre of investigation and analysis. For scholars working within this framework, state structures are the key to coming to grips with contemporary development processes.

Counter-revolutionist writers such as Bauer (1972), Bhagwati (1993), Johnson (1971) and Little (1982) attack the damaging consensus on development economics (Corbridge 1995: 7). Contrary to the opinion of the dependency school, the neo-counter-revolutionists argue that the developing countries are poor not because of the 'predatory activities of the First World and the international agencies that it controls but rather because of the heavy hand of the state and the corruption, inefficiency and lack of economic incentives that permeate the economies of developing nations' (Todaro 1997: 87). Central to such thinking are freer markets and correct pricing policies.

Recent theories and approaches

Most recently there has grown up a diverse group of 'new' approaches to development theory. Some of these include grassroots movements, the role of NGOs, gender issues, justice and democracy, citizenship and – most importantly – the environment.and local relationships. There is a tendency to shift emphasis away from large-scale theory to middle- or small-scale analyses – the local within the general development process. Again, some writers emphasise the 'development from below' approach or the 'basic needs' approach. There is also a movement which focuses on the 'empowerment of the poor' – emphasising people rather than places. And somewhat in sympathy with these movements is the New Growth Theory, which emphasises endogenous rather than exogenous growth. It provides

> a theoretical framework for analysing endogenous growth, persistent GNP growth that is determined by the system governing the production process rather than by forces outside that system. It restores the significant role of government policy in promoting long-run growth and development.
>
> (Todaro 1997: 93)

More fundamentally, the whole concept of development as a desirable process is under fire from some quarters in which development studies and theory are regarded as irrelevant. Anti-development ideas have been taken up by adherents from various sources. Development implies change, and all change can be for better or worse. Thus what might subjectively be termed 'good' development may be regarded as producing economic growth, national progress and modernisation along Western lines, and involving the provision of basic needs, sustained growth and improved governance. On the other hand, development might equally be regarded as 'bad' in that it involves a dependent and subordinate process, creates and widens spatial inequalities, harms local cultures and values, perpetuates poverty and poor working and living conditions, produces unsustainable environments, and infringes human rights and democracy. Environmentalists make up some members of the anti-development school, but some also come from those on the left who have seen the ineffectiveness of dependency thinking and have lost faith in socialism or communism. There are also those who argue that most development policies are imposed by the West, using Western concepts: 'development can be regarded as an attempt by the West to produce other societies in its own image' (Corbridge 1995: 8). And some of the most challenging work on development today draws on Said's account of orientalism (1979), Foucault (1979) and post-modernism to suggest that the modern concept of development is based on all sorts of colonising ambitions (Corbridge 1995: 8). Post-colonialism challenges the whole notion of the impact of imperialism on Third World cultures and opposes the ethnocentrism of white, Anglo-Saxon culture (Crush 1993).

Many other ideas and authors could be mentioned here, but the list is almost endless and it is reasonably easy to pursue the range of theories in a number of easily accessible sources referred to at the end of this chapter. Perhaps the last word on theories of development – an interdisciplinary field of study which can so easily seem complex and confusing – should go to Corbridge (1995) in his plea for tolerance and a recognition that there are many paths to development: 'Each of us holds hybrid views' (p. 19). He advises the student of development studies to be sceptical about

> panaceas for development thrown up by committed advocates of modernization theory, or neo-Marxism, or a populist anti-developmentalism. Development studies needs good theory, but it depends even more on a willingness to engage in reasoned argument and due consideration of other points of view.
>
> (Corbridge 1995: xv)

Summary

This chapter clarifies the approach used in this book and then deals with definitional problems surrounding the terms 'development' and 'the developing world'. It is pointed out that there is no accepted 'correct' definition of either of these terms, and that the reader should develop independent positions on the many points of discussion. The same goes for the discussion on theories of development.

Questions for discussion

1 What do you consider to be the most important criteria for determining a country's level of development?
2 Why should geographers be interested in and participate in the study of development?
3 Why do we need theory in development studies?

Further reading

The literature on the developing world is vast and interdisciplinary in its range but at this stage the following references will give a reasonable cover of the points discussed in the present chapter. They will also be found very useful throughout the rest of this book.

Corbridge, S. (ed.) (1995) *Development Studies: A Reader* London: Arnold.

Davis, D. (1992) 'Unlearning languages of development' *Latin American Research Review* 27: 151–68.

Dickenson, J. *et al.* (eds) (1996) *A Geography of the Third World* London: Routledge.

Potter, R., Binns, J., Smith, D. and Elliott, J. (eds) (1999) *Geographies of Development* London: Longman.

Todaro, M. (1997) *Economic Development* London: Longman.

The books by Dickenson *et al.* and Potter *et al.* are substantial, comprehensive texts written from very different points of view. The two approaches can be most usefully compared.

Mention should also be made of the companion volumes in the present series, many of which connect with this text at several points in the discussion and are referred to where appropriate.

2 ▶ Population and development

- Recent changes in the study of population and development
- Quantitative aspects – size, density, distribution, growth rates, life expectancy, sex ratios, age structure, migration, refugee movements
- Qualitative aspects – health, education, attitudes to work, human rights

Recent changes in the study of population and development

Over the past three decades there has been a significant change in emphasis in the debate on the relationship between population and development. Population as a resource, considered both quantitatively and qualitatively, is now less commonly regarded as *the* major problem in development thinking. In particular, the 'population problem', raising a neo-Malthusian spectre, no longer haunts every discussion. It has also come to be recognised that population is not so much the problem as the purpose of development. Development, however defined, is aimed ultimately at 'improving' people's lives in one way or another. There is also a perceptible shift in the balance between the large-scale, broad generalisations and the small-scale, local context of much of the current work in development studies (Corbridge 1995: 174); it is around individuals, households and small groups that much of the current debate is focused. This is especially logical in developing countries, where the extended family household is very important as the practical unit (Potter *et al*. 1999: 122). Nevertheless, just as there are many different perspectives on development, so there are wide differences of opinion about the role of population in development, and the following account attempts to identify the most important of these.

At the outset, however, it is important to emphasise that the data on which so much of the discussion on population and development is based are far from complete or reliable. Population censuses are costly and time-consuming, and depend for their success on such factors as levels of literacy and expertise in the conduct of population counts. Few of these requirements are present in most developing countries.

Quantitative aspects

Size

If size were any correlate of success, then clearly the economies of the developing world should be doing very well: of a world population of over 6 billion in the year 2000, some 84 per cent of people are in the developing world. The same point applies to country figures. If size had anything to do with economic success then China would be a paradise, containing as it does within its borders some 22 per cent of the world's population, and much the same could be said of India, with 15 per cent of the world's population; taken together, China and India have about half the total population of the developing world. Again, Nigeria, with 120 million, and easily the most populous country in Africa, would then be that continent's most prosperous nation. In fact, however, all these economies are well down the prosperity league table, whereas some – but by no means all – of the countries with the smallest populations have relatively prosperous economies. All the evidence indicates that there is no causal connection, either positive or negative, between population size and levels of development.

Density

As for population densities, it is perhaps worth considering whether it is purely fortuitous that levels of development – at least in the developing world – appear to be highest where population densities are highest. This applies equally at the continental level – where Africa, the least densely populated continent, is easily the poorest – and at the country level; thus Singapore, with one of the highest densities, is one of the wealthiest states. This whole question not only impinges on the debates about optimum population, overpopulation, underpopulation and population pressure, but is also relevant to the argument, first expressed some time ago by Boserup, that population pressure may be a necessary precondition for successful economic advance. Such a statement, of course, conflicts strongly with the opinion that 'population pressure' is a common cause of poverty – an opinion which ignores the fact that poverty or distress, relative to conditions in developed countries, are characteristic features of most developing countries everywhere, whether heavily populated or not.

Distribution

No attempt is made here to describe the distribution of population in any detail. Detailed maps giving the population distribution are available in most of the major texts, including those cited at the end of this chapter. Clearly the distribution – the result of a complex range of factors, varying from place to place – is very uneven, with the greatest concentrations to be found in the Asian sector and the smaller concentrations in sub-Saharan Africa. Of more immediate significance, perhaps, is the rural–urban distribution, though the figures for levels of urbanisation are exceptionally suspect. In the case of China, for instance, the most recent urbanisation figure is 50 per cent; but this makes sense only if set against the definition of 'urban' in China, which includes large areas which are rural but which have been placed under the jurisdiction of municipal or urban authorities. The same point explains the 100 per cent urbanisation figure for the city-state of Singapore.

Growth rates

Table 2.1 gives the growth rates of population and other data for selected countries. While in some countries migration can significantly affect population growth rates for short periods, it can legitimately be left out of any general discussion on population growth rates, for it is the balance between births and deaths – or natural increase – that most concerns us here. The figures given in Table 2.1 are particularly significant because they reveal how far along the demographic transition the various countries have moved. They are also interesting when compared with figures of two or three decades ago. Whereas at that time the highest rates of population growth were in Asia and Latin America, today the highest rates of growth are in sub-Saharan Africa and the Middle East.

What has brought about these changes? The decline in population growth rates has been greatest in China and parts of South and South-East Asia, and in Latin America. The main determinant here has been a decline in

Table 2.1 *Population, health and illiteracy data for selected countries*

Country	Population growth rates (1990–7 av. ann.)	Life expectancy at birth (1996)		Adult illiteracy rate: % of people over 15 (1995)	
		Male	Female	Male	Female
Bangladesh	1.6	57	59	51	74
Brazil	1.4	63	71	17	17
Chile	1.6	72	78	5	5
China	1.1	68	71	10	27
Eygpt	2.0	64	67	36	61
Ghana	2.7	57	61	24	47
Honduras	3.0	65	69	27	27
India	1.8	62	63	35	62
Indonesia	1.7	63	67	10	22
Kenya	2.6	57	60	14	30
Malaysia	2.3	70	74	11	22
Mexico	1.8	69	75	8	13
Nigeria	2.9	51	55	33	53
Pakistan	2.9	62	65	50	76
Philippines	2.3	64	68	5	6
Sierra Leone	2.5	35	38	55	82
Singapore	1.9	74	79	4	14
Tanzania	3.2	49	52	21	43
Zambia	2.8	44	45	14	29

Source: World Bank (1999)

the birth rates, but different factors have operated in different areas. Throughout the developing world population control measures are widely advocated, though with varying degrees of success. Fertility rates tend to be decreasing everywhere. In China the 'one-child-per-family' policy has been fairly effectively imposed, at least in urban areas, whereas in the Philippines the voluntary use of contraceptive methods has been less successful, having run into serious opposition from the Catholic Church. The same is true to some extent in Latin America, and in India coercion has had less effect than had been anticipated. The main problem now in China is whether the government should continue to tell women how many children they should have (*The Economist*, 21 November 1998: 21; Shen 1998). The earlier 'one-child-per-family' policy is beginning to break down, especially in rural areas and where a couple have already had a girl. But the control of birth rates has not occurred simply as a result of direct methods (i.e. contraception); indirect methods, such as education, have also had their effect, and there is now a general acceptance of the view that rising incomes are increasingly important in controlling population growth. In other words, development itself brings about lowered birth rates. This clearly affects the argument over whether population control is a necessary precondition for economic development, or whether it is development itself which leads to effective birth control. As for death rates, these have been lowered significantly in Asia and Latin America, reflecting the extent to which disease and malnutrition have been successfully controlled, though mortality rates may also be affected by sudden famines, floods, other natural disasters or wars.

Is it true that higher population growth rates (as distinct from large numbers or densities of population) tend to hold back development? Certainly it appears that those countries with the lowest rates of population growth are the most prosperous. While no simple correlation can be established, the economic consequences of rapid population growth are clear enough. The age structure is affected, leading to increases in the child dependency burden and to a rapid growth in the number of people of working age, bringing pressure to bear on governments to increase employment opportunities, to increasing levels of urbanisation and worsening problems of environmental control. High rates of population growth increase the danger that population growth rates will outstrip rates of economic growth, and so have negative effects on social and economic development generally. It is worth emphasising that the highest population growth rates today are in the poorest, most unstable and often in the most rapidly urbanising economies (Nixson 1996: 44).

Growth rates and the Malthusian spectre

The point has already been made that the 'population bomb' or 'population explosion' kind of thinking and writing that was so dominant a few decades ago has now lost much of its currency. It is significant, for instance, that at the Rio UNCED Conference in 1992, the countries of the developing world argued for a lower profile for population control in the final report, asserting that a high rate of population growth is a symptom, not a cause, of local environmental deterioration. 'Development is the best contraceptive' sums up what is now an influential viewpoint on this

Plate 2 *Mahatma Gandhi Road in Calcutta*
Photograph reproduced courtesy of Panos Pictures; © Heldur Netocny

whole question of population growth (Dickenson *et al.* 1996: 79–80). Another way of expressing the same point is: 'Poor people are not poor because they have large families. They have large families because they are poor' (Mandami 1972: 175).

Thus there is nowadays little support for the suggestion that the Malthusian spectre is still with us. It is not now universally accepted that high population growth rates are a prime cause of poverty. The problem is poverty, not population. Moreover, one hears very little today of 'overpopulation' or 'underpopulation'. The developed world now seems to talk less of limiting population growth and more of providing aid for development. There are many factors that have brought this about. In Asia and Latin America women's empowerment has helped to bring down birth rates, as it has in the developed world; direct action through contraception, and indirect action through education have helped; and in China, the one-child-per-family policy has had its impact. As development proceeds, too, there is not the same demand for children to provide insurance and labour. Much remains to be done, notably in sub-Saharan Africa, but fears about the future of population are now concentrated more on such issues as the age structure and on the potential conflicts between mutually hostile groups. Significantly, the 1994 Cairo Conference UN World Plan of Action recommended a well-balanced and practical policy that most developing countries were ready to sign up to.

On the other hand, population growth policies are sometimes directed at ends above and beyond the desire for economic development or the need for insurance against old age and labour provided by children. Thus in Malaysia there have been selective policies designed to encourage the Malays to 'go for five' in an attempt to catch up with the numbers of Chinese in Malaysia and in neighbouring states. And in Singapore, selective policies have been designed to encourage the relative growth of the educated and professional classes. Nevertheless, there remains the problem of unemployment, especially as job creation becomes increasingly difficult in a period of rapid technological change where fewer jobs per unit of output are required.

Life expectancy

Reflecting as it does the effect of a number of variables, life expectancy is regarded as a useful measure of levels of development; it will be remembered that longevity, measured by life expectancy at birth, is one of the criteria used to determine the Human Development Index discussed in Chapter 1. Table 2.1 gives the figures for selected countries over a period of years, and these figures do reflect very well changes in levels of development in different parts of the world. Life expectancy averages 52 in least developed countries, 61 in other developing countries and 75 in developed countries.

For Africans living south of the Sahara, average life expectancy is a mere 51 years, some 10 years less than for people anywhere else. In part this grim figure is a result of the spread of AIDS and HIV, which the World Bank claims infects between a tenth and a quarter of adults in eighteen African countries. Sub-Saharan Africans also have the highest infant mortality rates, the lowest levels of school enrolment and the least access to safe water. In the decade 1990–9 their economic growth was the lowest of any region, though the situation was hardly less dire in the former Soviet Union and countries of Eastern Europe.

Sex ratio

The significance of the sex ratio, which continues to slightly favour females, has changed somewhat over the past decades as women are increasingly gaining equality of opportunity and status within many societies. Women now seek a more direct share in the labour market, and their economic status is being enhanced, for example in the new electronics industry in the Asian newly industrialising countries (NICs) (Dickenson et al. 1996: 62).

While the sex ratio in China for first births is about average, the surplus of boys increases for every subsequent birth, and the same phenomenon has been observed in South Korea and Taiwan. New techniques of determining sex – an important factor in Chinese societies – includes amniocentesis in Korea and Taiwan; and mainly ultrasound in China. The resultant dearth of daughters will help create the world's biggest group of frustrated bachelors. According to recent estimates, by 2020 the surplus of Chinese males in China will exceed the entire female population of Taiwan. Furthermore, the effect of this surplus of males will be aggravated by a dramatic fall in fertility to below replacement levels – possibly down to under 1 in some places, such as Shanghai (*The Economist*, 19 December 1999: 98).

Age structure

It is now widely, if rather belatedly, accepted that one of the most important demographic implications of reduced rates of natural increase in many countries throughout the world is the effect on the age structure. It is unfortunate – though it was always predictable – that success in dealing with the so-called 'population explosion' has resulted in dangers arising from top-heavy age structures. As developed countries, such as the United States, the United Kingdom and Japan, are now discovering, an ageing population is a characteristic of countries with low rates of population growth, leading to increasing levels of elderly dependency, a problem which must be faced by all countries, as Peterson has shown (Box 2.1). At present (the year 2000), 16 per cent of China's population

---

Box 2.1

Global ageing

Global ageing is inevitable. By 2030 there will be only 1.5 working taxpayers for every non-working pensioner in the developed world, today the ratio is 3 to 1. And, as the case of China shows, the same trend is now occurring in many developing countries. Increased standards of living and advances in medicine have contributed to the trend, as have the 'baby boom' (resulting in an imminent bulge in the retirement population in developed countries) and decreasing fertility almost everywhere. Throughout the world there is a striking decline in fertility rates, which are down from 2.7 to a level approaching the replacement rate of 2.1. The developing world, with the exception of much of sub-Saharan Africa, is also experiencing this drop in fertility, and the populations of some developing countries are now ageing faster those of a typical developed country. At the state level this can lead to a demand for labour immigration; in some European countries, indeed, non-Europeans already make up 10 per cent of the population (*Foreign Affairs* January/February 1999, 78, 1: 42–55).

Global ageing means higher pension and health care costs, together with changing economic structures. A World Bank Report *Averting the Old Age Crisis*, published in 1994, spells out what is likely to happen and how the problems might be met or averted. But solving these problems – by greater personal savings and investment, and by later retirement – faces formidable difficulties even in developed countries. In poor, developing countries the difficulties would seem to be insurmountable.

Source: *The Economist*, 13 February 1999, and Peterson (1999)

are aged 60 or over. Many people worry that there are too many people in China – 'the Yellow Peril' – but the real worry today is that in future there may be too few people in work in China to support the rapidly growing population of retirement age. This problem, referred to by some as the 'demographic implosion', is present everywhere but is potentially serious for many developing countries (Figure 2.1).

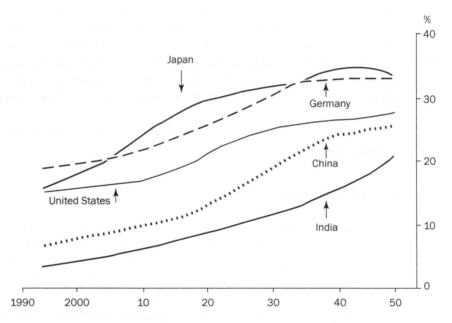

Figure 2.1 *Projected percentage of population aged 60 or over*
Source: United Nations, *The Economist*, 21 November 1998

In much of China, skilled workers are tasting poverty and insecurity for the first time in a generation. The baby-boomers (especially those born in 1950–5) are now bearing the brunt of economic restructuring: 'China's economic reforms will be at the cost of a whole generation', according to Peng Xizhe, Director of the Institute of Population Research at Fudan University. The baby-boomers have already had their share of misfortune: a large age-group with relatively poor health and poor education, having suffered during the Cultural Revolution. In China at least 30–55 million workers are not needed. Many baby-boom couples will find themselves in old age with only one child to support them. 'They pose the most intractable problem anywhere in the world.' There has been a rapid rise in pensioners supported by diminishing numbers of workers in present and former state enterprises. The World Bank says,

'China will have a high-income economy's old-age burden with a middle-income economy's resources for shouldering it' (*The Economist*, 21 November 1998: 81–2).

Migration

Of the three variables (births, deaths, migration) that can directly affect population numbers, migration is normally of relatively little long-term significance, though it can have locally important effects, as has been shown in sub-Saharan Africa, where trans-state migration, resulting from wars and natural disasters of one kind or another, has become a powerful factor in African demography. Otherwise, migration is important in any discussion on development because it may affect significantly the composition of the population of a country and have a noticeable impact on the economy. This point is expanded in a brief discussion of the effect of the Chinese diaspora in South-East Asia (Box 2.2). But many other examples could be referred to, including the South Asian (Indian) diaspora, the migration of Mexicans and other Latin Americans into the United States, and the current migrations of people from Eastern to Western Europe – that is, from the former Soviet Union, Poland, Hungary, the Czech Republic, Romania and Bulgaria.

As for the causes of international migration, there are many theories about these. Referring to the migration from Eastern Europe to Western Europe, Austria's International Centre for Migration Policy Department estimates that at least 400 000 people are now smuggled into the European Union each year and that, whatever the real reasons for migrating, they often try to claim asylum as refugees. However, the common distinction drawn between 'push' and 'pull' factors is not always easy to demonstrate in practice; and the suggested dominance of the economic factor is equally not always easy to prove.

Refugees

Refugee problems may arise from floods, droughts, earthquakes, volcanic activity, hurricanes and tidal waves as well as from wars or 'ethnic cleansing'. In Shanghai, fertility plummeted in the late 1960s, and the one-child-per-family policy had by then been in place for a long time. Life expectancy there has risen by 33 years in five decades, and the

Box 2.2

The Chinese diaspora

In this brief summary of the migration of Chinese people out of China into countries of South-East Asia, it must not be forgotten that the Chinese have also moved elsewhere, more particularly to the western United States and Canada, and to Western Europe. As Winchester has expressed it, the movement of Chinese overseas represents 'the greatest diaspora in the world' (1991: 220).

As far as the Overseas Chinese in South-East Asia are concerned, Table 2.2 shows the number of Chinese in five of the states in the region. In Indonesia there are over 4 million Chinese, or about 3 per cent of the population. In Malaysia there are rather more, almost 5 million Chinese, but this gives them a much higher percentage of 30–5 per cent of the total population. Brunei has a Chinese percentage of almost 30 per cent and in the Philippines fewer than a million Chinese represent a percentage of the total population of about 1.5 per cent. Singapore has easily the largest percentage (76 per cent) of Chinese. In the five countries there are altogether some 12 million Chinese, though the total number of Chinese in South-East Asia as a whole, including Hong Kong, Taiwan and Vietnam, is about 40 million.

Table 2.2 *Numbers of Overseas Chinese in South-East Asian countries*

Country	Ethnic Chinese	Total population	% Chinese
Indonesia	4 116 000	147 000 000	2.8
Philippines	699 000	46 000 000	1.5
Malaysia	4 882 300	12 736 637	30.9
Singapore	2 038 000	2 413 945	75.9
Brunei	85 000	300 000	28.0

Source: Various, but especially Rigg (1991: 110) and Suryadinata (1989: 1). From Hodder (1992: 44)

Concerning the motives for this great migration, all the evidence suggests that the primary motive was economic. Most of the migrants came from the heavily populated provinces of south-eastern China – Guangdong, Fujian and Guangxi – where poverty and recurrent famines were common. There were clear economic forces tending to push the Chinese out of their homelands. But there was also a good deal of internal political strife and physical insecurity, and it is a matter of interpretation whether the economic 'push' motive was as clear-cut as is commonly suggested. There were certainly very strong economic 'pull' factors encouraging the Chinese to look overseas for a better life.

Our interest here is the effect these Overseas Chinese have had on development in the various countries of the region. There is little doubt that this effect has been considerable. Some would argue that the Overseas Chinese in South-East Asia have played and continue to play a crucial role in providing the catalyst for economic growth in the region. As Redding (1990) has pointed out, the 40 million or so Overseas Chinese in the area have between them a GNP two-thirds the size of the 1.2 billion Chinese in China. In Peninsular Malaysia the Chinese own and control nearly 40 per cent of the corporate sector and their contribution to economic development has been critical. In Sarawak, too, where most of the non-indigenous people are Chinese, they occupy the more densely populated coastal and lower valley strips in the relatively well-developed west. The Chinese, too, are the main urban people in Sarawak, as they are throughout most of Malaysia: 60 per cent of the population of Kuching, the capital of Sarawak, are Chinese. Throughout Malaysia the Chinese dominate in business, and it is generally true that the whole internal trade of the country passes at some stage or other through the hands of Chinese merchants or middlemen.

The same is largely true of Indonesia, where Chinese dominate in most commercial sectors. Chinese operate especially at the higher levels of trade, capital-intensive and high-technology trade being mostly in Chinese hands. Property markets are also dominated by the Chinese; Chinese private firms are growing rapidly; and Chinese companies dominate in shipping. In the Philippines, Chinese own some 40 per cent of the total assets of private domestic banks.

Clearly the Chinese are an important element in the economic life of these South-East Asian countries. Explanations of their success are many and varied: 'the reasons used to explain the success of the Chinese in the region include their high motivation (which is linked to their status as migrants), their Confucian work ethic and business acumen, and the role of Chinese business networks' (Rigg 1991: 111). They also include the suggestion that the Chinese happened to be in the right place at the right time; and that they had the skills, the temperament, drive, experience and urbanised culture that coincided with the needs of the European colonialists. Moreover, the Chinese are said to be hard-working, frugal, patient and willing to make less profit per unit of sale. Their work ethic is certainly very strong and sometimes frighteningly uncompromising. The stigma attached to play and relaxation is also very apparent, especially among the older Chinese. It is also pointed out that the Chinese have a propensity to invest accumulated capital in a trade in which the entrepreneur has acquired certain basic skills through previous apprenticeship or in-service training as an employee.

That the Chinese have made an important, even critical, contribution to economic growth in Singapore, Malaysia, Brunei, Indonesia and the Philippines is undeniable. 'As entrepreneurs, in commerce and in business, the Overseas Chinese continue to act as the catalyst of economic growth. They have been described as the omnipresent, inescapable part of the warp and woof of every country between Vietnam and India, Luzon and Timor' (Winchester 1991: 237).

Source: Hodder (1992; 1995)

number of deaths has sometimes exceeded the number of births of official residents. But illegal immigrants began to pour into Shanghai and now account for over one-fifth of Shanghai's population.

However, as Figure 2.2 illustrates, the scale and scope of the problems arising from major refugee and internally displaced peoples worldwide are now very great. And since April 1999, when Figure 2.2 was drawn, other major movements, notably in Kosovo and East Timor, have occurred.

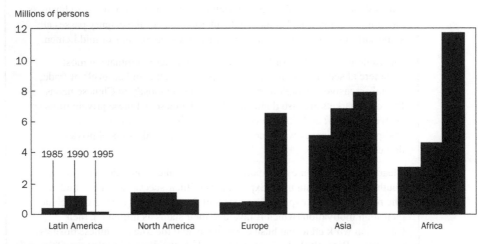

Figure 2.2 *Refugees in different regions*
Source: World Bank (1999), from date of the UNHCR, 1995

The history of the past fifty years, and the 1990s in particular, has been the forced exodus of populations from their homes and great waves of people fleeing state-inspired terrorism, ethnic cleansing and human rights abuses. The office of the United Nations High Commissioner for Refugees (UNHCR) estimates that there are now over 25 million people who have been forced to leave their countries and another 25 million who are now internally displaced within their own countries but unable to return to their lands or villages. These refugees are referred to as the Fourth World, living without rights, home or hope of return.

But official figures are just the tip of the iceberg. And in an increasingly globalised world where populations are burgeoning in the poorest countries, abject poverty has caused millions of economic and environmental refugees to flee the land. Internal displacement of populations is now as significant in humanitarian terms as the more conventional flights across state boundaries. And an increasing characteristic of the worldwide refugee phenomenon is that few refugees return for good.

In the past few decades the movements outside Europe have been much greater. The great diasporas of the twentieth century include the Jews and Armenians fleeing genocide, massive forced relocations under Stalin, and millions of people fleeing communism; and more than twenty countries have been torn apart by populations fleeing repression.

The lot of the refugee or the displaced person is increasingly hard. Globally there is mounting rejection of refugees, and states have been quick to erect physical and administrative barriers to their movements. And, as Collins (1995) has shown for Belize, efforts to deal with refugee resettlement can easily be misconceived or misdirected.

Qualitative aspects

From a development point of view it could be argued that of far more importance than numbers in any attempt to explain differential levels of development between different countries is the quality of the population. It may be that an important part of the answer lies not in a country's endowment of natural resources, nor in the size of its population, but rather in the energies, skills and organisation of its people. This point has already been referred to briefly in the previous section on the Chinese in South-East Asia, and other cultural determinants will be looked at in Chapter 3. But here we must look at some of the basic indicators of 'quality', especially health and education.

Health

Table 2.3 describes health in selected countries of the developing world according to several indices: access to clean water, access to sanitation and infant mortality rates. Other indices sometimes referred to are population per nursing person, percentage of births attended by health staff, and percentage of babies with low birth weights. Of the available figures – and there are some serious gaps – Indonesia and the Philippines come off badly, while in most respects Singapore, Hong Kong and Japan do well. As for infant mortality rates, these are remarkably low for Japan,

Table 2.3 *Some health indicators for selected countries*

Country	Access to safe water (% of population)	Access to sanitation (% of population)	Infant mortality rates per 1 000 live births
Angola	32	16	124
Bangladesh	79	35	77
Brazil	72	41	76
Egypt	64	11	53
Ghana	56	27	71
Honduras	65	62	44
Malawi	45	53	133
Nigeria	39	36	78
Sierra Leone	34	11	174
Singapore	100	97	4
Uganda	34	57	99
Zimbabwe	74	58	56

Source: World Bank (1999)

Hong Kong and Singapore. There are rather lower figures for daily calorie supply per capita in Vietnam, Indonesia and the Philippines than elsewhere, but these figures do not reveal anything significant in the absence of data for the quality of diet. By standards of world comparison, none of these indicators is particularly disturbing, though a good deal remains to be achieved in certain countries.

However, the problem of food supply in developing countries is by no means solved. It is true that the rate of food production has outstripped the rate of population growth in most areas, with the important exception of sub-Saharan Africa; and the Green Revolution technology since 1960 has made more food intake possible in South Asia. Nevertheless, it is easy to compile a list of facts which reveal how much remains to be done in the developing world: 1 billion people living on diets with calorie deficiency; 1 billion living with no access to health services; 1.3 billion living with no access to safe drinking water; 1.9 billion living with no sanitation; 192 million children malnourished. The average number of doctors per 100 000 people is 4.8 in the least developed countries, 210 in developed countries; and most health facilities are in cities, out of reach of the majority of the population of developing countries (Dickenson *et al.* 1996: 68).

There is also the problem of AIDS/HIV, notably in Southern Africa. In Botswana, Malawi, Uganda, Zambia and Zimbabwe, AIDS is now considered to be the leading cause of death between the ages of 15 and 39. HIV in sub-Saharan Africa is mainly spread through heterosexual intercourse and perinatal transmission, which can occur *in utero*, during delivery or after birth through breast milk. This issue needs to be kept in perspective, of course. Over sub-Saharan Africa as a whole, the incidence of AIDS is 12.4 per thousand (less than the figure of 22.7 for the United States), though the incidence is as high as 96.7 in Zimbabwe. But whatever the incidence, it has serious effects on life expectancy, infant mortality, population growth rates, costs of health care, and the productive capacity of the economies.

But what relevance has all this for development prospects, however these prospects are defined or measured? To some extent, certainly, better health and nutrition may result directly from development. Nevertheless, and quite apart from all moral and humanitarian arguments, the literature does suggest a positive effect of good health and nutrition upon productivity, especially in agriculture. Increases in work productivity, and in a child's ability to learn, have also been correlated with good health. Studies in China over many years, for instance, have shown that protein–energy malnutrition is related to low cognitive test scores and poor school performance. And in Indonesia a study found that iodine deficiency reduced cognitive performance among 9- to 12-year-old children.

Education

Education is represented in Table 2.4 by the percentage of age groups enrolled in primary and secondary education in selected countries. Differences between countries are least marked in the primary sector and only just apparent in the secondary sector. Adult illiteracy rates are often high and gender related as well as varying dramatically both within and between countries (Table 2.1). These are predictable differences, but they demonstrate how important education is to a rapidly developing country. Recent research also points to a very strong link between primary school education and economic growth. Studies in Malaysia and Kenya confirm that schooling substantially raises farmers' productivity. Literacy levels average 45 per cent in least-developed countries and range to 64 per cent in other developing countries, but it is important to recognise that such average figures ignore key exceptions such as Cuba, which may have important lessons for development strategies. Investment in education is

Table 2.4 *Educational indicators for selected countries*

Country	Primary	Secondary
Brazil	90	19
Costa Rica	92	43
Egypt	89	65
Indonesia	97	42
Malawi	100	66
Namibia	92	36
Nicaragua	83	27
Philippines	100	60
Venezuela	88	20
Zambia	77	16

Source: World Bank (1999)
Note: The figures show the percentages of the relevant age groups enrolled in primary and secondary education

clearly worthwhile for many reasons, including the fact that educated farmers are more likely to adopt new technologies and to produce higher returns on their land. And it is not just in the labour market that improvements seem to result from education. It is estimated that one year's education for women can lead to a 9 per cent decrease in under-5 mortality. Other things being equal, the children of better-educated mothers tend to be healthier (Figure 2.3).

On the other hand, 'education' is a broad, rather vague term covering many types and qualities of training. It may be questioned whether education in its narrowest academic sense is necessarily beneficial to development in developing countries, however desirable it might be to an individual in a cultural sense, or to economic progress in a technologically advanced developed country. Education for politicisation, to instil an ideology, to encourage instant obedience and to discourage individualism and criticism of authority – is this either desirable or necessary? And does not academic education in the Western sense encourage elitism and a tendency to despise commercial, agricultural and industrial occupations? A good deal of literature focuses on these questions and on the precise form education should take in any particular country, and it seems that few worthwhile generalisations are possible. It has been pointed out that the role of education and the forms it takes will be very different for a dispersed, rural, Muslim population in Indonesia, for a largely Chinese population in the prosperous city-state of Singapore, and for the peasantry of the Communist-controlled society of

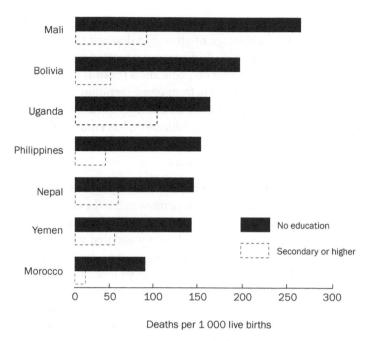

Figure 2.3 *Child mortality by educational attainment of the mother*
Source: Macro International, various years, after World Bank (1999)
Note: Mortality data are for children under 5

China. It is important that education should not create disaffected intellectuals; but it is also important that it should be in close touch with the life of the community as a whole. Education should be in harmony with the technical and administrative requirements of each country. This means, for example, that short-term technical training, without the luxury of a lengthy academic or more formal education, may in certain circumstances be the most appropriate form of education (Hodder 1992: 21).

The India case

India's poverty-reducing strategies focus increased public spending on expanding the poor's access to *quality* education and health care, funding these increases by reducing subsidies elsewhere in the economy. It is accepted that the returns on primary education are particularly great: better family health, smaller family size, and healthier children for educated women, for instance. A mother's primary education may do even more than food subsidies to improve child nutrition. Educated workers can take advantage of growth in demand for labour to raise their productivity and earnings. And because educated parents are more likely to send their children to school, education (even if limited to the primary level) perpetuates the benefits from one generation to the next. It is no coincidence that Kerala State, with its development perspective and its strong political commitment to education, has enjoyed the highest levels of both male and female literacy and the fastest decline in the incidence of poverty over the past forty years.

If India is to meet its objectives of sustaining high rates of economic growth with equity, then schooling must reach the economically and socially disadvantaged. This will require policies that can expand both the quantity and the quality of schooling and that work to eliminate social exclusion based on income, gender or caste. The problems are concentrated in the states of Bihar, Madhya Pradesh, Orissa, Rajasthan, Uttar Pradesh and West Bengal. Notwithstanding the remarkable accomplishments of the District Primary Education Project (DPEP) in increasing enrolment in lower primary education, three-quarters of the 33 million 6- to 10-year-olds in these seven states are still not in school.

The policy prescriptions for bringing the poorer children into school are not obvious. Success will depend on much stronger political commitment to educate the children of the poor. There is no question that spending levels on education inputs need to increase. But empirical findings suggest that increased spending alone will not be enough to improve enrolment and attainment of the poor. Improvement in quality of schooling is crucial to attract poor children to school and keep them there. Because the opportunity cost is high for them and their families, poor children will not go to school if they perceive it as a waste of time.

Improving quality of education, however, needs fundamental reforms to change the incentive framework within which teachers, school officials, bureaucrats and politicians operate. These reforms include decentralisation of control over the provision to the local areas themselves, direct parental involvement, competition through school choice, and community involvement. But raising quality is not as easy as increasing spending, and information on what has worked elsewhere in the world and within India is needed to inform public policy about the design of an effective educational system.

Effective health programmes must complement education in raising the potential productivity of labour to reduce poverty. At present, public expenditure on health has only a limited redistributive impact. Despite its relatively equitable distribution, it is small relative to overall health use, and its effects on health outcomes appear to differ greatly from one intervention to another and from place to place, especially in its impact on the poor. There are four priority areas for increasing the impacts of public spending on the health of the poor, and indeed on the economy in general.

First, combating communicable disease and expanding public health interventions would deliver substantial gains from public health spending, particularly for the poor. Second, improving access to safe water sources and sanitation facilities and vaccinations would help reduce infant and child mortality and thus reduce fertility and improve maternal health. Because these are activities in which the poor are vastly underserved relative to the non-poor, public interventions in these areas would achieve the biggest impact on the poor. The net cost to the government of extending water and sanitation facilities to poor areas may not be very large, since willingness to pay for these goods is usually quite high and could cover the extension of the system. Third, analyses in India have shown that health education concerning basic hygiene, the value of better nutrition, and preventive care such as public campaigns against tobacco use and for the use of appropriate measures to avoid contracting HIV/AIDS and other sexually transmitted diseases is an important part of encouraging behavioural changes needed for long-term improvements in health outcomes. Fourth, because the rural poor must often meet the financial burden of medical emergencies through debt, distress sale of real assets, or reductions in food or other important consumption items, there is merit in subsidising hospital treatment. The benefit of providing this 'in effect' social insurance – on top of the value of the service itself – is in the range of 40 to 70 per cent of the costs of providing the service to patients in the lowest 40 per cent of the population – people for whom insurance is not a realistic option. Public subsidies to hospital care can thus play an important redistributive role as long as referral systems are reformed to ensure that access is based on need rather than income and social status (World Bank 1998b).

The Africa case

In Africa, however, more and more children do not go to school at all. According to UNICEF, about 40 million children in sub-Saharan Africa get no basic teaching. This is due largely to lack of resources, itself commonly a reflection of huge debt burdens. It can also reflect the incidence of AIDS. It is frequently suggested that putting into primary education some of the money now spent on debt servicing, on armies and on higher education – combined with changes in attitudes to girls' education – could quickly give most African children basic skills in reading and writing. As Sen puts it, 'educate part of a community and the whole of it benefits' (1994: 5). He believes that education and learning drive economic growth.

Developed countries, such as the United Kingdom and Japan, provide convincing proof of the critical importance of school education. In Japan, at the beginning of the Meiji Restoration in 1868 only about 15 per cent of the population were literate. In 1872, however, elementary education was made compulsory throughout Japan, and later, in Japan's colonies of Korea and Taiwan, education and literacy were make central planks of development planning,

Attitudes to work

Other qualities are often mentioned but there is not a great deal of hard information to go on. This is true, for instance, of attitudes and aptitudes regarding work. As we have noted above, some writers refer to the work ethic, or its absence, among particular groups of people. Apart from the climatic determinism argument already referred to, it has been suggested that racial or ethnic differences can explain why, for instance, the Chinese in Malaya seem so much more thrusting and successful than the indigenous Malays. But is this difference innate? Or is it the expression of some cultural trait, such as religion or sets of values?

This is looked at briefly in Chapter 3, but it might be useful here to suggest that entrepreneurial ability and energy, initiative, independence and an enthusiastic work ethic might reflect not the race, religion or culture of an individual or group, but rather an institutional framework which allows an individual's energies, initiative and ambitions to flourish and operate freely. Does the case of the Chinese suggest that it is the social, political and economic institutional framework in mainland China – state control, collectivism and the whole paraphernalia of Chinese communism – which was, at least until very recently, responsible for China's heavily bureaucratic and inefficient economic performance? After all, the highly motivated and successful Chinese in Hong Kong,

Taiwan and Singapore all came from China in the first place, thereby casting doubt on any 'innate' characteristic as an explanation of success.

Much of this discussion is based on uneven and dubious evidence about the relationship between human resources and development. But no kind of simple causal connection can be made. Even when one is discussing the quality of human resources, can it really be suggested that there are differential capabilities that make some some groups more effective than others in the development process? Yet it is clearly going too far to suggest that natural and human resources are rarely critical and may even be irrelevant factors in development. After all, each government, each economy, has to operate within its own resource environments, its own particular combination of natural resources and, as Chapter 3 shows, its own cultural legacy.

Human rights

It is difficult to know exactly where to place any discussion of human rights. The topic has already been raised in the earlier discussion on refugees, but some authors interpret the term to include reference to the fact that many, if not most, people in the developing countries are not provided with 'basic needs' or 'human rights' in terms of adequate health care, education or housing (Potter *et al.* 1999). In other words, human rights are most lacking where poverty is most marked. This broad definition of human rights is of course legitimate, but more commonly the term is used to refer to circumstances in which populations are subjected to cruelty, torture, expulsion and repression. There is an assumption that human rights are best protected in formally democratic states and that the growth of democratic governments will lead to better human rights records. As Nowak and Swinehart (1989) have shown, the percentage of formally democratic states in the world increased from 25 per cent in 1973 to 68 per cent in 1993.

Certainly the human rights records of many states have been and remain appalling, but it is difficult to know what, apart from international pressure, other governments can do about this. The Vienna Conference on Human Rights in 1993 focused particularly on certain elements of populations: women, religious groups and other minority groups. Other governments have to decide whether to maintain a dialogue or to hold back aid or grants of various kinds – for almost all offending countries consider human rights to be strictly their own affair.

Summary

This chapter examines the changing relationship between population and development, dealing first with the essential quantitative aspects of population, from size, density and growth rates to migration – a topic illustrated by the case of the Chinese diaspora, which has clearly had a very significant impact on the scale and nature of development in the region. There follows a discussion on the qualitative aspects of population as a human resource in the development process. This section is dominated by health and educational considerations but there is increasing concern today on the issue of human rights. Finally, no firm causal connections can be established between the various elements of population and existing levels of development, but the discussion does demonstrate how much needs to be done to improve the quality of life among the majority of people in the developing world.

Questions for discussion

1 Is the rate of growth of population the major problem facing developing countries in trying to achieve development?
2 Why is an ageing population a serious problem worldwide?
3 With reference to specific cases, comment on the causes of international migration.

Further reading

In addition to the general texts listed at the end of the previous chapter, the following will be found useful:

Barrett, H. and O'Hare, G. (1992) 'India counts its people' *Geography* 77: 170–4.

Boserup, E. (1965) *The Conditions of Agricultural Growth* London: Allen and Unwin.

Chant, S. (ed.) (1992) *Gender and Migration in Developing Countries* London: Belhaven.

Demeny, P. and McNicoll, G. (1998) *Population and Development* London: Earthscan.

Hettne, B. (1990) *Development Theory and the Three Worlds* London: Longman.

Hodder, R.N.W. (1992) *The West Pacific Rim* London: Belhaven.

Skeldon, R. (1990) *Population Mobility in Developing Countries* London: Belhaven.

World Bank (1999) *World Development Report, 1998/99* Oxford: Oxford University Press.

3 ► Culture and development

- Views on the role of culture
- Ethnicity
- Elites and class
- Religion and codes of behaviour
- Gender
- Corruption

Views on the role of culture

Culture is not an easy concept to pin down; nor is it easy to define. In his companion volume in this series, *Cultural Geography*, Craig (1998) defines cultures as 'sets of beliefs or values that give meaning to ways of life and produce (and are reproduced through) material and symbolic forms'. He further emphasises the value of taking a relativistic stance over culture, pointing out that it applies to all societies: 'Cultures are not only about exotic faraway peoples, but also about the way we, in the West, do things.' This is an important point, supporting as it does the dangers of stereotyped views of developing countries, ignoring the many similarities that exist between all cultures, and even casting doubt upon whether the concept of a 'western' culture is in any real sense valid. Another writer defines culture as 'the binding element that ties individuals together through their integrated patterns of behaviour, and as such it acts to include some individuals within the group in question and to exclude others from it' (O'Malley 1988: 328). Clearly, any discussion of culture can range widely over many aspects of human activity, but in this chapter we are concerned primarily with the relationship between culture and development, especially economic development.

There are strongly divergent views about the role of culture in development. Some writers, including many economists, argue that culture plays no positive role at all in the development process, though it may have a negative role in the sense that it can hamper or inhibits economic development, or prevent it from occurring at all. Moreover, the role of culture in development is an immensely broad and complex topic, without clear disciplinary boundaries, so that no one person can possibly be qualified enough to analyse convincingly the economics *and* the culture, as well as all the links in between, if indeed there are any such links. Cultural explanations are also dismissed by some writers as reflecting an inability to explain development in any other way: 'for those who are baffled by the plethora of conflicting explanations and contradictory evidence, there is refuge in the cultural factor' (Riedel 1988: 2).

An opposing vewpoint argues that culture is an important, perhaps *the* critical factor in development. Writing of East Asia, for instance, Gourevitch argues that

> the expression of cultures is overwhelming. We find in this region one of the oldest civilisations known, China, and some of the youngest. With such richness, the possibility of explaining important phenomena through the causality of culture seems irresistibly tempting. . . . Culture has been mobilised to explain many if not all of the phenomena of great interest to us.
>
> (Gourevitch 1989: 11)

While it is admitted that culture is not an easy concept to discuss, one author argues that 'to ignore culture largely on the grounds that it is awkward to deal with is both intellectually unsatisfying and potentially costly' (O'Malley 1988: 327).

The standpoint taken in this book is to lean more to the first than to the second of these approaches: to take the view that while cultural elements do often assist in the understanding of development as a process, and particularly in certain specific problems, these cultural elements cannot be used as valid explanations for different levels of economic success. Cultural 'explanations' of why people think and behave as they do, and why they organise their economies and political structures as they do, are of little help to us in our attempts to understand why one economy has been successful while another has not:

> analyses which dig deep into a country's philosophy, religion, customs, magic and the broad sweep of its history in an attempt to explain why people . . . come to think, behave and organise in the way they do are of course fascinating. The cultural characteristics they reveal may indeed help or hinder development. But they cannot explain or contribute towards an explanation of that development
>
> (Hodder 1992: 41)

Without in any way relegating culture to 'the dustbin of developmental economics' (Riedel 1988: 26), a good deal of evidence suggests that we should examine critically any crude cultural determinism. It has been suggested elsewhere that economic growth derives from an overpowering determination to achieve economic growth; that a conscious decision is made to pursue such growth; and that any cultural factors are simply used or rejected as seems appropriate.

However, this viewpoint is very generalised and it is important that every reader should make up his or her own mind concerning the role of culture in development and, if possible, examine the relative success or failure of economies in the context of globalisation. It may be, of course, that when different cultural elements are examined separately and are related to specific situations or circumstances, then rather different conclusions will be reached. For the rest of this chapter, therefore, we look at specific elements of culture and relate them to specific examples.

Ethnicity

Alhough there are many different definitions of an ethnic group, all of them include the notion of a group with some kind of corporate identity or allegiance. Members of an ethnic group tend to feel a loyalty to the idea of their group's political independence, whether actual or sought; they are also likely to express loyalty to a common culture which parallels or transcends loyalty to a state. As for ethnicity, this refers to the sense of identification, psychological commitment, historical membership, or set of values. Ethnicity is to be found everywhere, in both developed and developing countries, past and present. In the United Kingdom, for instance, is it legitimate to regard the 'Welsh' as a 'tribe' or 'ethnic group'?

However, an element of interaction is believed to be essential to ethnicity, and the degree of ethnicity depends on the degree of interaction between one ethnic group and another (Box 3.1). Ethnicity, unlike an ethnic group, cannot exist in isolation. And it seems a valid principle that the more homogeneous the population, the more stable the economy. Examples commonly cited are Singapore, Taiwan and South Korea.

Ethnic and racial disparities are serious in many areas of human development (UNDP 1998). In South Africa whites had a life expectancy of 68 years in the early 1990s, 14 years more than the 54 years for

Box 3.1

The Hausa and the Western Ibo

The case of the Hausa and Western Ibo, and their different reactions to living in the same Yoruba city (Ibadan, a city of over 1 million population) in south-west Nigeria, provides an interesting case study of the importance of culture in a multi-ethnic situation. Here in the indigenous Yoruba city of Ibadan the Muslim Hausa (from northern Nigeria) and the Western Ibo (from the lands east of Yorubaland) are two important non-indigenous groups, and in the early 1920s the British colonial powers allocated them specific residential districts in the city. Since that time both groups have been subjected to the same forces, pressures and opportunities. But today the two groups present marked contrasts. The Hausa have preserved, indeed deepened, their cultural distinctiveness: for the most part they still speak only Hausa and interact socially only among themselves. A cleavage, sometimes a tense cleavage, exists between them and the host Yoruba community in Ibadan. The Western Ibo, on the other hand, have almost lost their cultural distinctiveness over the same period. Their residential segregation has completely broken down and their compounds are occupied by people from different ethnic groups. They did have a group association – the Western Ibo Union of Ibadan – but it met infrequently. Like many other tribal associations in Africa, the Western Ibo Union did not aim at any exclusive ethnic policy but was concerned solely to promote the successful adaptation of its members to modern urban conditions. Whereas everything within the Hausa group was aimed at maintaining the distinctiveness of the Hausa, among the Western Ibo exactly the opposite occurred. The Western Ibo are now largely indistinguishable from the Yoruba majority. The second-generation Western Ibo speak Yoruba and often cannot speak Ibo at all, and they, unlike the Hausa, frequently intermarry with the Yoruba and other groups.

The Western Ibo, then, have reacted to outside pressures by adjustment, assimilation and adaptation; the Hausa, however, have reacted by reorganizing and redefining their own traditional customs or have developed new customs under traditional symbols. Why this should be so is an interesting question that cannot be pursued here. Some would explain it in terms of inherent differences in character and temperament between the thrusting, ambitious Ibo and the conservative, traditional Hausa. But such an explanation is probably far too simplistic.

Source: Cohen (1976)

blacks. In Malaysia the incidence of income poverty among ethnic Malays, at 24 per cent, is nearly four times that among the ethnic Chinese, at 6 per cent (Box 3.2). In Canada 35 per cent of Inuit men are unemployed, compared with 10 per cent of other Canadian men. In the

United States 31 per cent of Hispanics aged 25–65 have not completed ninth grade, compared with only 6 per cent of whites. This uneven progress in human development over the years and the existence of a significant backlog of human deprivation have not only resulted in persisting disparities, but have generated forces that are reversing human progress in several areas (ibid.).

Ethnicity, then, can have serious negative effects. In Yugoslavia 'ethnic cleansing' has created distress and poverty on a grand scale, as it has in Rwanda (the Hutu and Tutsi). In Rwanda the Tutsi account for only a small proportion of Rwanda's population but they provide the most powerful elements in the government. And in Nigeria the mutual antagonisms between the Muslim Hausa and Fulani in the north and the Yoruba and Ibo in the south have made any sense of national unity or economic progress peculiarly difficult to achieve. Since being created by the British, Nigeria has been divided between north and south, between religions, between language groups (250 or so), between rich and poor, and between civilians and the military. In Nigeria's main oil-producing areas in the south in the Niger Delta, the Ijaw, Ogoni and other communities of the delta have issued warnings that unless they see benefits from the oil, no more oil will be allowed to be pumped from their land. Little has been done by the oil companies or the central government to reduce pollution or develop the delta, where fishermen say that their catches of fish have plummeted, from producing a commercial surplus to being below what they need to subsist.

In South-East Asia ethnicity is expressed in many ways – perhaps most notably in the contact between Chinese and indigenous peoples in the manner discussed in the previous chapter (Box 2.2). But ethnicity is at the heart of the problems of many beleagured minorities, including the Melanesians of Irian Jaya, the H'Mong of Laos, the Dayaks and Iban of Borneo, the non-Burmese population of Myanmar, and the Madurese in Indonesia. The Madurese now make up 8 per cent of the 4 million population of West Kalimantan. They moved to Kalimantan in response to the Indonesian government's policy of relieving population pressure in Java by developing the Outer Islands; but conflict between the Madurese and the Indonesian Malays has recently exploded into serious ethnic riots.

On the other hand, ethnicity can play a largely constructive role in the creation of national unity. For after all, is not ethnicity just one form of allegiance? Ethnicity can also be an asset in modernisation in that it

---

Box 3.2

The Malays

The history of the Malay people of Malaysia, Indonesia and the Philippines is one of diversity and constant movement and change, having been influenced by the Chinese and many other groups, including the Indians. Indian Islam took hold on the Malay Peninsula and by the fifteenth century had spread throughout Indonesia to the Sulu archipelago and western Mindanao in the Philippines. The particular brand of Islam that emerged had already to some extent become Indianised, and was in any case a tolerant and adaptable philosophy which had no difficulty in accepting and integrating with traditional beliefs and customs. It soon became and has remained the dominant religion or philosophy among the Malays in Malaysia and Indonesia, and in parts of the southern Philippines.

The Malays are essentially an island and peninsular people, and this has made for cohesion at local levels and has reinforced the strength of the family, though rather less so than in the Confucian cultures to the north. Furthermore, in the predominantly Muslim world of the Malay peoples, an individual's performance in life, and place in death, is held to be of more significance than the relationship between superior and subordinate persons. Some writers view such a statement as a crude rationalisation of Malay people's relaxed attitude to work and material achievement. But whatever the truth of the matter, it clearly affects their potential for rapid economic growth and has been incorporated into several other interpretations, such as that of Boeke. His view on the indigenous peoples of Indonesia (formerly the Dutch East Indies) was that the indigenous, colonised peoples there possessed cultural values that were largely inimical to economic growth. This, he argued, was one reason why the territory developed what he called 'dualism': one part of the economy involved the energetic, economically successful colonists (Europeans), together with a few exceptional individuals, including some Chinese; the other part was formed by an indigenous population, the bulk of whom were held back from participating effectively in the process of economic growth by their own cultural values and expectations.

This kind of stereotyping was of course very characteristic of European colonialism and was frequently used to explain differentials in levels of development between territories and ethnic groups.

Source: Hodder (1992)

---

enlarges the sense of collective identity and security at different levels and can maintain links between town and country. Members of a particular ethnic group coming to a town may, initially at least, develop their links with the social network of their own home areas. And reference has already been made in Chapter 2 to the positive role that the Chinese have played in economic development throughout South-East Asia.

A rather different approach to this whole issue is to argue that the ethnic group, as a focus for corporate identity and common loyalty, may well be a positive force for good; but that ethnicity, or the force that derives from conflict and tends to be defensive and parochial, is undesirable and nothing less than a particularly virulent form of nationalism.

It has been noted that ethnicity as a term and as a subject of study is of recent origin. The term 'ethnicity' first appeared in the 1950s. 'It has now become the major source of social and political conflict in both developing and developed societies' (Hutchinson and Smith 1996: 4). But though the term itself is recent, the sense of kinship, group solidarity and common culture to which it refers is as old as the historical record. Ethnic communities have been present in every period and continent, and have played an important role in all societies.

For at least 150 years liberals and socialists confidently expected the demise of ethnic, racial and national ties and the unification of the world through international trade and mass communications. But in fact in Europe and in the Americas ethnic movements unexpectedly surfaced in the 1960s and 1970s, in Africa and Asia they have been gaining force since the 1950s, and the demise of the former Soviet Union has encouraged ethnic conflicts and national movements to flourish throughout its territory. Since 1990, twenty new states based largely upon dominant ethnic communities have been recognised.

The validity of all such generalisations, however, must take into account the great variety of conditions and circumstances in different countries. This point can be illustrated by a few examples. In Latin America there are very significant differences between ethnicity in Mexico, Peru and Bolivia. In 1970 only 7.8 per cent of the Mexican population were defined as Indian, and these Indians were divided into fifty different linguistic groups. Geographically and politically, the Indian population tends to inhabit the periphery. The central Nahua zone is not considered to be Indian. Mestizos occupy the centre and regard the Indians as peasants and rural poor. The term 'Indian' serves only as an instrument of class oppression. In the case of Peru, however, Indians account for about 47 per cent of the total population and are concentrated in the southern highlands. Dualisms are apparent in the fact that (i) the Indians occupy the highlands, the whites and mestizos the coast; and (ii) the whites and mestizos occupy the cities, the Indians the countryside. With white help the mestizos clearly dominate the Indians. In Bolivia, by contrast, the distributional pattern of the various groups is very dispersed. There is a variety of regional and ethnic fragments, rather than a highland–coast or urban–rural division.

Racial, ethnic and religious discrimination in Latin America can be pernicious. The indigenous Indian populations have lagged behind other groups on almost every measure of economic and social progress. Thus 90 per cent of Guatemala's native population are poor, compared with 50 per cent for the rest of the population. And this poverty particularly affects women.

Another case study is taken from Hungary and its Gypsies, where the Gypsy (or Roma) population numbers some 500 000, or nearly 6 per cent of the population. They regard their plight as being much worse than that of black people in the United States. Fifty per cent of the Gypsies have no job, compared with 10 per cent of black people in the United States; many live in the north-east, an area seriously affected by the closure of decrepit plants, and their life expectancy is only 65. The Hungarian government agrees that the problem of the Gypsies is serious. But it argues that the problem is not that the Gypsies are victims of ethnic hatred, but that they are poorly educated and cannot compete for jobs. Leaders of the Gypsies, however, disagree and point out that their plight in Hungary is similar to that of Gypsies in other parts of Central Europe.

Elites and class

An elite may be defined as a group which influences power and clearly redefines the norms of a society. It can also be defined as a group which enjoys a position of pre-eminence over others which is shown by various acts of deference towards its members. The elite should also have some degree of corporateness and exclusiveness, with definite barriers to admission.

In the context of the present study, the major role of elites is in stimulating the growth of the economy, and this role has been likened to the role played by entrepreneurs in the industrialisation of North-Western Europe in the nineteenth century. Elite groups are in touch with international trading networks and, as a result of the prestige they command over the mass of the population, they help to diffuse new ideas and ways of life throughout a country. Through the intermediaries of the elites, new ideas may seem attractive and worth emulating, and without this in-between group with its distinctive role, the diffusion of ideas would probably be much slower and less widespread. In other words, the existence of an elite makes economic development more possible, social integration easier and modernisation more feasible.

Cohen illustrates this point by drawing on an analogy between the Hausa in Nigeria, especially in the southern city of Ibadan, with the business elite of the City of London. In the City of London the elite

> speak the same language and presumably partake in the same culture of the wider society, but when one looks closely into their style of life one will discover subtle peculiarities – in accent, manner of linguistic expression, style of dress, patterns of friendship and of marriage, etiquette, manners – that are organizationally instrumental in developing boundaries, communication, and other mechanisms for the organization of the group. The elite thus coordinate their corporate activities through their style of life.
>
> (Cohen 1976: 46)

Cohen transfers this notion to the Hausa traders in Ibadan. A Hausa dealer from northern Nigeria will entrust his money and goods in the south only to a Hausa broker; and a Hausa in Ibadan is anxious to preserve the symbols of Hausaism, not only by dressing, speaking and behaving like a Hausa, but also by separating himself physically or residentially from the rest of the population (Box 3.1).

How can we evaluate elitism? The negative features are considerable and obvious. Elites are socially divisive. Where they are associated with a particular ethnic group they tend to emphasise and reinforce the distinctiveness and ethnic basis of life and activity, and to reinforce the power of ethnic minorities as well as majorities. Intellectual elites frequently become dangerously alienated from the major practical problems of economic, social and political development – an alienation which is only too apparent in many universities in the developing world. Further, in countries where political problems are not yet very manifest, the problem of creating an educated elite with no jobs – unemployed white-collar workers – is becoming acute. On the other hand, the positive functions of elitism cannot be ignored. Elitism can function as a mechanism for modernisation, affecting motivation and incentives, and linking up within countries and linking countries together.

Religion and codes of behaviour

Some of the early influential work on the relationship between religion and development was provided by Weber, who suggested that the greatest drive to industrialise in Europe came in areas located far from Catholic orthodoxy. More recently Vogel has suggested that the great drive for East Asian industrialisation in recent years has prospered in areas where trade and commerce were most highly developed – that is, in the centres of traditional Confucian orthodoxy.

Opinions differ very widely about the role religion can play as a factor, either positive or negative, in development. As for Confucianism, many writers consider this to be the key to understanding the economic success of East Asia. Thus, in Singapore the economic success of the Chinese there has been explained in terms of Confucianism, emphasising such qualities as a belief in hard work and filial piety. Mirroring Weber's 1958 Protestant ethic argument for Western Europe at the beginning of the Industrial Revolution, it is argued that Confucianism 'instils values such as hard work, thrift, goal attainment, and family loyalty, among others, that prepare people for effective behvaiour in market economies' (Gourevitch 1989: 1). And Downton suggests that societies with Confucian-Buddhist roots 'are proving more effective with the industrial and technological challenge on the eve of the twenty-first century than are . . . those countries with a predominantly Christian-Hebraic tradition' (quoted in Winchester 1991: 61); similar comments have been made about the Muslim world (Box 3.2). On the other hand, such views seem to ignore two facts. One is that it was only 40 years or so ago that Confucianism was widely regarded as an inhibiting factor explaining the *lack* of economic progress in East Asia. The other is that, on a global scale, it is precisely those societies and economies with a 'predominantly Christian-Hebraic tradition' which have made the most progress in meeting 'the industrial and technological challenge on the eve of the twenty-first century'.

Gender

In recent years there has been a great increase in the literature on women and their role in development in different cultures. Gender is now incorporated into a Gender-related Development Index (GDI) to facilitate comparisons between countries and regions (UNDP 1998). The GDI uses the same variables as the HDI, referred to on page 8. The difference is that the GDI adjusts the average achievement of each country in life expectancy, educational attainment and income between women and men. The GDI also adjusts the maximum and minimum values for life expectancy, to account for the fact that women tend to live longer than men. For women the maximum value is 87.5 years and the mimimum value 27.5 years. For men the corresponding values are 82.5 and 22.5 years. The greater the gender disparity in human development, the lower the country's GDI relative to its HDI. If one looks at the tables compiled in the UNDP (1998), it is clear that the human development achievements of women fall below those of men in all 163 countries covered by the table, and the shortfall in the GDI relative to the HDI reflects this inequality. For several countries the GDI rank falls short of the HDI rank by 20 points or more: Oman, Saudi Arabia, the Islamic Republic of Iran, the Syrian Arab Republic, Algeria, Libya and the United Arab Emirates. In these cases the place of women in Muslim society is clearly the major factor.

Another measure of increasing use is the Gender Empowerment Measure (GEM), which uses variables constructed explicitly to measure the relative empowerment of women and men in political and economic spheres of activity (UNDP 1998). The first two variables are chosen to reflect economic participation and decision-making power: women's and men's percentage shares of administrative and managerial positions and their percentage shares of professional and technical jobs. The third variable, women's and men's percentage shares of parliamentary seats, is chosen to reflect political participation and decision-making power.

Progress in building women's capabilities has been significant but there is a serious delay in creating real opportunities for women. The lack of equal opportunies for women to participate in economic and political life is partly captured by the GEM, estimated for 102 countries. At the top of the GEM rankings are three Nordic countries: Sweden, Norway and Denmark. And some developing countries do even better than developed industrial countries – thus Trinidad and Tobago and Barbados are ahead of the United Kingdom and Ireland; Cuba and Costa Rica are ahead of France and Israel; and China and Mexico are ahead of Japan.

The evidence provided by the GDI and GEM data in the latest UNDP annual report (1998) illustrates that societies have made some progress over the past thirty years or so in achieving a more equitable distribution of the benefits of development between women and men. Gender gaps in education and health have narrowed rapidly. Female life expectancy has increased 20 per cent faster than male life expectancy over the past two decades. Education levels have been steadily rising for women in developing countries. The gaps between women and men in adult literacy and school enrolment were halved between 1970 and 1990. In primary schools the enrolment levels for girls, once 74 per cent of those for boys, are now about 90 per cent on average.

In India there are moves to reserve one-third of all seats in the national parliament and in state assemblies for women. To quota-happy India, reservations seem a natural cure for inequalities; India already reserves some government jobs for members of lower castes. India and China are unusual in having rather more men than women. Yet only 43 per cent of Indian women are literate, compared with 69 per cent of men.

Culture of corruption

Until recently it was believed that although corruption and/or cronyism were characteristic of many countries, both developed and developing, this had no marked effect on the development process. There has also been a good deal of tolerance of 'good' as against 'bad' corruption. Most recently, however, corruption has been shown to be a significant restrictive factor and responsible for a great deal of poverty in all countries, including, for instance, the United Kingdom and Russia.

In the 1990s alone, governments in Italy, Brazil, Pakistan and Zaire have fallen partly because the people they governed would no longer tolerate the corruption of politicians. The same applies to Ferdinand Marcos in the Philippines, deposed in 1986. Corruption, according to some writers, is partly caused by a corrupt form of capitalism which diverts resources from economically well-founded enterprises to those that were merely well connected.

And now Mauro, an International Monetary Fund (IMF) economist, has shown that countries with high levels of corruption have less of their GDP going into investment and have lower growth rates. 'An increase in the corruption level from that of Singapore to that of Mexico is equivalent to raising the tax rate by over 20 percentage points' (Mauro 1999). In Albania businesses pay in bribes an average of 8 per cent of their turnover, or about one-third of their potential profits. In Indonesia *pungli* adds up to about one-fifth of total operating costs. And in Nigeria, Abacha damaged the country's oil refineries so much that the world's

	0	2	4	6	8	10
Denmark	-------------------------
Sweden	------------------------
Canada	-----------------------
Singapore	-----------------------
Netherlands	----------------------
Australia	---------------------
Britain	---------------------
Germany	--------------------
Hong Kong	--------------------
United States	--------------------
Israel	-------------------
Chile	-------------------
France	-------------------
Portugal	------------------
Spain	-----------------
Japan	-----------------
Belgium	----------------
Malaysia	----------------
Taiwan	----------------
South Africa	---------------
Hungary	---------------
Greece	---------------
Czech republic	--------------
Italy	-------------
Poland	-------------
South Korea	------------
Brazil	-----------
China	-----------
Turkey	-----------
Mexico	----------
Philippines	----------
Argentina	----------
Thailand	---------
India	--------
Russia	-------
Venezuela	-------
Colombia	------
Indonesia	------
Nigeria	------
Cameroon	-----

Figure 3.1 *Corruption perceptions index (10 = least corrupt)*

Source: Transparency International, *The Economist*, 3 October 1998

eleventh biggest oil exporter found itself importing fuel. Oil earns 90 per cent of Nigeria's foreign exchange and provides 70 per cent of its GDP, which is why the Ijaw, Ogoni and other Delta tribes want some pay-off. The culture of corruption is endemic and powerful in Nigeria. Successive military governments have kept themselves in power through a system of institutionalised theft that has inculcated the habit in many institutions and all echelons of society.

Figure 3.1 shows the corruption index for selected countries and illustrates that there is no clear division between the most corrupt and the least corrupt in terms of developed and developing countries. The case of Bulgaria (Box 3.3) is also significant.

This culture of corruption is difficult to remove, wherever it occurs, but some coordinated attempts are now being made. Western-based multinational companies are now required by Western governments to work together to limit the amount of bribery engaged in. International financial institutions such as the World Bank and the IMF are increasingly linking the aid they offer to poor countries to those countries' willingness to provide 'good governance'.

How space-specific is culture? And is there any fundamental difference between the culture of developed and that of developing worlds? The evidence presented here throws some doubt on this, although it is true that every culture is in some respects distinctive. But in the context of development geography it is easy to ignore or underestimate the similarities between countries. And this is relevant in any discussion of the concept of globalisation of culture, as addressed in Chapter 8.

Box 3.3

Corruption in Bulgaria

As in many other ex-communist countries, factory bosses in Bulgaria saw the advent of the free market as an opportunity for plunder. They paid too much for raw materials, and charged too little for finished goods. To cover their losses, they borrowed from the banks. The government of barely reformed communists, itself up to its neck in doubtful business dealings, winked at the stripping of the banks. But then ordinary Bulgarians, who could see where this was leading, started pulling money out of the banks. The central bank, which was simultaneously battling a loss of confidence in the currency, created liquidity to bail out the banks, then soaked it up to prop up the currency. Disaster was averted only because the government called an election in April 1997, and lost it; the newly elected reformers made a corruption crackdown one of their top priorities.

Source: *The Economist*, 16 January 1999: 28

Summary

The discussion centres first on the relationship between culture and development, presenting the opposing viewpoints on whether culture can explain different levels of economic success. The bulk of the chapter, however, looks at specific elements of culture – ethnicity, elites, class, religion, gender and corruption – and relates them to specific examples. Although every culture is clearly unique in some ways, the evidence presented here suggests that all cultures possess many striking similarities. Any form of cultural determinism should be critically examined.

Questions for discussion

1 How would you define culture in the context of development geography?
2 Give examples of 'the culture of corruption' in your own country. How far do you believe this hinders economic growth?
3 Critically examine reasons why it is important to accelerate the closing of the gender gap if development is not to be held back.

Further reading

Bryant, R. (1996) 'Romancing colonial forestry: the discourse of forestry as progress in British Burma' *Geographical Journal* 162: 169–78.

Cohen, A. (ed.) (1974) *Urban Ethnicity* London: Tavistock.

Craig, M. (1998) *Cultural Geography* London: Routledge.

Hodder, R. (1992) *The West Pacific Rim* London: Belhaven.

Hutchinson, J. and Smith, A.D. (eds) (1996) *Ethnicity* Oxford: Oxford University Press.

Lloyd, P. (1979) *Slums of Hope? Shanty Towns of the Third World* Harmondsworth: Penguin.

Mills, C. (1996) 'Gender and colonial space' *Gender, Place and Culture* 3: 125–47.

4 Rural-agricultural development

- Urban bias and technological change
- Types of agriculture
- Problems of transition to commercial farming: land reform, improved and alternative technologies
- The rational peasant
- Women's roles
- Environmental issues
- Rural development
- Agriculture or industry?
- Peri-urban vegetable gardening in China

Urban bias and technological change

Most people in the developing world live in rural areas and are engaged, directly or indirectly, in some form of agriculture; and it is also in these rural areas that the greatest number of the poorest are to be found. As Table 4.1 reveals, there is some variation in this pattern, with Latin America having a rather lower percentage of people living in rural areas compared with Asia and, more particularly, with Africa. And yet there is in the literature – and, indeed, in practical development planning policy – a tendency to accept these facts almost as 'given' and to focus attention on non-agricultural, non-rural and urban-industrial problems

Table 4.1 *Percentage of population living in rural areas*

	1980	1997
World	60	54
Low- and middle-income countries	68	60
East Asia and Pacific	79	67
Europe and Central Asia	42	32
Latin America and the Caribbean	35	26
Middle East and North Africa	52	42
South Asia	78	73
Sub-Saharan Africa	77	68
High-income countries	25	22

Source: World Bank (1999)

of development. This 'urban bias', as it is termed, has characterised much writing and policy, both during and since the colonial period, and still exists in most developing countries.

More specifically, Chambers (1983: 13–23) has identified six biases which, he argues, impede our contact with and understanding of rural life and poverty in the developing world. These are:

1 *spatial biases*: urban, tarmac and roadside, in which visitors, officials and researchers confine their experience largely to the towns, roads and roadsides and fail to experience the real rural poverty and problems they need to identify;

2 *project bias*: this emphasises the showpieces in development, which are often only tiny atypical islands of activity of little relevance to the rural poor;

3 *person bias*: the bias here is in favour of the rural 'elite' (articulate and fluent, but not necessarily with the best interests of the poorest at heart); men, rather than women; users and adopters; and those who are active and present in front of the official, visitor or researcher;

4 *dry-season bias*: the bias here is in favour of visiting rural areas in the dry season, when travel is more comfortable, rather than in the wet season, when most farming activity goes on and when the problems of rural development are at their most stark;

5 *diplomatic biases*: politeness and timidity: the rural poor may be excessively polite and timid about meeting the visitor, official or researcher, or ashamed to take them into the poorest parts of a village; and

6 *professional biases*: professional training, values and interests sometimes focus on the less poor – on the 'progressive' farmers, for example. And 'specialists' may be so immersed in their specialism that they do not embrace the whole problem of rural development and rural poverty.

To some extent these biases are understandable. Governments usually gain most of their support in the urban areas rather than in rural areas where populations are more difficult to access and the people more remote from the centres of power and decision-making. There is often, too, a good deal of ignorance in governments about the rural poor, and what their needs and problems really are. The result is to marginalise the rural poor majority and so to perpetuate their isolation and poverty.

The implications of this, not only for development, but also for the provision of basic food supplies, are serious, even leaving aside for the moment the impact of famines, floods and other natural disasters on food production of the kind experienced in Honduras on 1998 (Box 4.1). Food consumption per capita is low, and over the past two decades has actually declined in some areas, most notably in Africa. Moreover, even where cash non-food crops are involved, failure to improve production for export has had dramatic effects on economic prosperity. For all kinds of reasons, too, failure to deal adequately with the rural-agricultural sector has its impact on all other forms of development and needs to be addressed as quite fundamental to all other sectors.

At its simplest, improvement in the rural-agricultural sector can be said to imply a technological transformation, from low-productive and

---

Box 4.1

Hurricane Mitch in Honduras

The destruction wrought by Hurricane Mitch in Honduras in late 1998 highlights the fragility of the agricultural economies in many developing countries. Its effects were most damaging in Honduras, though it also affected Nicaragua, Guatemala, El Salvador and south-eastern Mexico. Early estimates put Honduras's crop losses at 70 per cent and the number of homeless at 2 million out of a total population of only 6.3 million. The small, largely subsistence farmers, with only a hectare or so of land, grow staples such as beans, maize and rice. Beans, ready to be picked, were especially badly hit, though there are normally two bean harvests a year. Maize was heavily hit, too, but only one-fifth of the expected maize crop was lost overall. As for commercial crops – bananas, coffee and cane sugar – Honduras lost about half of its bananas and almost a quarter of its coffee; taken together these two crops were responsible for Honduras losing one-eighth of its exports.

But crop losses were not the only serious effect of the hurricane. When rivers that had burst their banks receded they left sand up to 2 metres deep. Some farmers with fields next to rivers lost them for good. In the hills, torrents of water washed soil away, or rinsed the nutrients away, thereby reducing fertility.

This was bad news for a country whose food output has been rising more slowly than its population: in 1996 Honduras grew 27 per cent less food per person than in 1980. But it is much worse news for individual farmers, since many were barely scratching a living anyway. The same goes for workers on the big banana plantations. Overall, in the plantations Hurricane Mitch will result in temporary or permanent unemployment for thousands of workers.

Source: *The Economist*, 20 February 1999: 67–8

labour-intensive peasant agriculture to high-productive, capital-intensive commercial farming, in which agriculture progresses with fewer farmers. In general the productivity gap between farming in the developed and developing worlds is said to be increasing, though this is less true of East Asia than of other parts of the developing world.

Some criticism has been directed at the above statement about the process needed to raise rural-agricultural standards, to produce more food and non-food crops, and to bring about development in its broadest sense. 'Bottom-up' and 'peasant-focused' approaches to rural agricultural development still have their protagonists. In South Africa, for example, Nel *et al.* (1997) have described the 'bottom-up' rural development strategy; this has been applied in the light of the failure of so many 'top-down' schemes and is now a successful community-driven development initiative. But accepting as valid the statement in the previous paragraph about the aims and purpose of rural-agricultural development, we look first at the range of types of agriculture and then examine the problems and policies involved in making the transition from simple peasant agriculture to efficient commercial farming of the kind found in most developed countries.

Types of agriculture

The problem of how to classify types of agriculture in the developing world has attracted a large amount of literature and many conflicting viewpoints. For the purposes of this discussion, however, a simple threefold classification will be used to demonstrate the main problems facing the increase of agricultural production.

Subsistence agriculture

Subsistence agriculture is generally thought of as the most common form of agriculture in the developing world, involving family farming of a small acreage, producing primarily food crops for a family's own 'subsistence', using largely family labour, and working with the simplest of tools and techniques. Where land supply conditions make it possible, the clearance of land may not be permanent, and shifting (or shifting-field) cultivation is practised, but this is now much less widespread than it used to be. The same can be said of pure 'subsistence agriculture', in which no crops are produced for cash exchange at all; this is now a very rare phenomenon. More commonly, a farmer will practise a system which may involve a considerable degree of subsistence agriculture, varying perhaps from year to year and from season to season, yet always involving an element of cash cropping. Development of subsistence agriculture implies higher and/or more secure living standards, higher levels of efficiency, and thus greater specialisation.

Mixed farming

Mixed farming is, strictly speaking, a system in which crop and animal production are combined to form systems of mixed farming and alternate husbandry, producing both crops and livestock simultaneously or alternately from the same land. This form of farming is more common in Asia than elsewhere in the developing world. But even in Africa, nomadic tribes may for short periods pasture their herds on the resting lands of other cultivators – a practice which helps maintain fertility and may assist the cultivators to shorten the fallow period. However, mixed farming in developing countries rarely goes beyond a system in which animals simply supply manure, energy for pulling farm implements, and milk or meat. One important type of this kind is wet rice cultivation, a system associated with some of the most densely populated and long-settled parts of Asia and the homes of a number of long-established civilisations. It is permanent-field cultivation *par excellence* and commonly involves the use of draught animals for power in pulling ploughs and in fertilising the land. It is now being extended into many other developing regions – notably into the Amazon Basin and in West Africa

Commercial farming

Whereas the dominant interest in peasant farming – either subsistence or mixed farming – is producing enough food for the farmer, in commercial farming the dominant interest is in financial profit. Commercial farming is normally associated with the large, specialised farm or plantation for such crops as bananas, cocoa, coconut products, coffee, oil-palm products, rubber, sugar and tea. But in some parts of the world the smallholding system is used for certain crops: cocoa (in West Africa) and rubber (in parts of Malaysia). Smallholder production of cash crops is widely underestimated, especially in such exporting countries as Myanmar and Thailand (rice), Colombia, Côte d'Ivoire and Togo (coffee), Bangladesh (rice and jute), Ghana and Nigeria (cocoa), and Uganda (cotton). Smallholdings – to use the term in its widest sense to cover small shifting-field cultivation as well as the most sophisticated forms of peasant permanent-field cultivation – have always been responsible for the bulk of agricultural production, if not of large-scale export production. And the role of the smallholder in commercial agriculture is increasing, though without entirely displacing the plantation system.

The plantation system refers to a large-scale combined agricultural and industrial enterprise that is both labour intensive and capital intensive; it also raises, and usually processes industrially, agricultural commodities for the world market. Size and processing equipment are often important criteria, and large-scale enterprises may ultimately become agribusinesses.

In the simplest peasant system of agriculture, the two major factors of production used are land and labour. But elsewhere, increasingly sophisticated and capital-intensive techniques are used to create what is in effect 'industrialised' agriculture. To state this point more precisely, the development of agriculture from one stage to another can be summarised as follows:

1 peasant subsistence and shifting-field cultivation: labour-using; capital-saving;
2 peasant permanent-field cultivation: labour-using; capital-saving;
3 the beginnings of 'industrialised agriculture' (smallholder or plantation) – application of crop research, pesticides, fertilisers, mechanisation, roads, credit facilities: increasingly labour-saving and capital-intensive;
4 'industrialised agriculture'.

It is how to move as rapidly as possible from 1 to 4 that presents a main purpose of agricultural development policy. But in this process a number of critical problems arise.

Problems of transition from peasant to commercial farming

> Unless low-productivity peasant agriculture can be transformed rapidly
> into higher-productivity farming in Latin America and Asia (primarily
> through judicious land reform accompanied by concomitant structural
> changes in socio-economic institutions) and Africa (basically through
> improved farming practices and greater price incentives), the hundreds of
> millions of impoverished and increasingly landless rural dwellers face an
> even more precarious existence in the years immediately ahead.
>
> (Todaro 1997: 314)

Land reform

The issue of land reform helps to explain some of the differences and the
different nature of the problems in agriculture between Latin America,
Asia and Africa referred to by Todaro in the above quotation. And, as
Potter *et al.* argue,

> patterns of landholding in poor agrarian economies are key determinants
> of 'agrarian structure', which concerns the different ways in which land
> and labour are combined in varying forms of production, as well as the
> social relations, such as class, which 'structure' the processes of
> production and reproduction.
>
> (1999: 272)

Latin America

In Latin America the dominant form of land tenure grew out of the
colonial period and is characterised by what is known as the latifundio–
minifundio pattern in which a large amount of land is owned by a very
small number of landowners. Some authorities identify four groups:
minifundios, which provide employment for a single family (up to 2
persons); family farms, which provide employment for 2–4 persons;
medium-sized farms, which provide employment for 4–12 persons; and
latifundios, which provide employment for 12 or more persons. But the
critical point here is how these figures relate to the amount of land
occupied by each of these four groups.

To take the case of Colombia, 5 per cent of the land is occupied by
minifundios, 25 per cent by family farms, 25 per cent by medium-sized

Plate 3 *Woman migrant worker carrying a load of cotton she has just picked on a foreign-owned estate near Santa Cruz, Bolivia*

Photograph reproduced courtesy of Panos Pictures; © Sean Sprague

Plate 4 Quechua Indian women working a plot of idle land in an occupied hacienda in Ecuador
Photograph reproduced courtesy of Panos Pictures; © Julio Etchart

farms and 45 per cent by latifundios. At the same time, 66 per cent of the value of agricultural production is produced by minifundios and family farms, whereas the medium-sized farms and latifundios account for only 34 per cent of the value of agricultural production. And as far as labour is concerned, the minifundios account for 58 per cent of the labour employed, whereas latifundios account for only 4 per cent of the labour employed: the latifundio system covers large plantation estates as well as cattle ranches.

These figures reveal that the greatest intensity, and indeed efficiency, of agricultural production is in the minifundios and family farms; indeed, large areas of some latifundios are not even cultivated, for the wealthy landowners of the latifundios may maintain their holdings not for their actual or potential contribution to agricultural output in the country, but rather for the considerable power and prestige they bring to their owners.

The distribution of land for agriculture in Latin America is clearly very unequal indeed, and this fact no doubt helps to explain the popularity of the neo-Marxist dependency writing particularly associated with that continent. Some writers argue that inequalities are getting worse – a process that can be related to ideas of the 'dynamic' of development. According to the FAO, 1.3 per cent of landowners in Latin America hold 71.6 per cent of the entire area of land under cultivation. It is only in a few countries, notably Mexico, Bolivia and Cuba, that drastic land reforms have taken place in recent years.

Asia

In Asia the basic problem which needs to be addressed in any attempt to assist the technological transformation in agriculture is one of overcrowding and excessive fragmentation of landholdings. As Todaro has put it,

> if the major agrarian problem of Latin America can be identified as too much land under the control of too few people, the basic problem in Asia is one of too many people crowded on to too little land. For example, the per capita availability of arable land in India, China and Japan is 0.29, 0.20 and 0.07 hectares, respectively.
>
> (1997: 298)

As a result of European colonial rule over much of South and South-East Asia, European land tenure systems involve private property ownership backed up by the law. This has meant in practice that the informal, traditional village structure of rights and obligations, in which the allocation of land was determined by village heads in a sensitive manner reflecting needs and disasters, was removed. A further effect has been to encourge absentee landlords who let out their lands to sharecroppers and tenant farmers. This has inevitably led to heavy borrowing and to the growth of a moneylending culture. It is partly for this reason that Asian peasant cultivators have seen their economic status deteriorate steadily over time.

In Asia, of course, population growth has in the past been a prime factor in creating the fragmentation and parcelling of land, with all its ill effects, but more recently the rates of population growth have dropped – in some areas quite markedly. As we noted in Chapter 2, it is now in Africa that one finds the highest population growth rates, together with some of population growth's most serious implications.

Africa

In Africa the agricultural systems differ, in some cases very widely, from those found in Latin America and Asia. Although plantations are found in selected areas, the commonest form of farming is the family farm in which subsistence (though increasingly rarely 'pure' subsistence) farming is dominant. In these circumstances the village land is held by the village, and the village head plays much the same role as used to occur in Indian farms before the European colonial period: informally allocating land according to need, and providing access to water and land for all members of the village community.

The major problem in Africa is the very low productivity of the typical African family farm, where only the simplest hand tools are available and where the use of animals is often made impossible because of diseases caused by such insects as the tsetse fly. Shifting cultivation, too, is still characteristic of some areas, and labour demands vary quite dramatically from place to place, year to year and from season to season, resulting in labour demand shortages and surpluses. The result of all these factors is a constantly low level of agricultural output, a situation made more serious by the increasingly rapid population growth in the continent. Shifting cultivation is now breaking down in most areas, and owner-occupier plots are being subjected to pressure from population and from the growth of towns, the penetration of the money economy, soil

Plate 5 *Woman shifting cultivator using a simple hoe in central Nigeria*
Photograph: B.W. Hodder

Plate 6 Ewe fisherman in Togo

Photograph: B.W. Hodder

erosion, deforestation of marginal lands, and the introduction of land taxes. Mixed and modern commercial farming is becoming more common. As we noted earlier, Africa has suffered most from its inability to expand food production at a pace sufficient to keep up with its rapid population growth. And, significantly, Africa's dependence on food imports, notably wheat and rice, is increasing rapidly.

Improved and alternative technologies

Many of the technological changes involved in productivity improvements are exactly the same as those introduced into Europe and other developed countries over the past century or so. These include mechanisation; application of the results of research into crop improvements, disease-resistant and drought-resistant varieties; the application of appropriate fertilisers; the control of pests, diseases and weeds; and the application of research into plant populations and plant spacing. Each of these has its problems, including particularly the fact that it requires the use of capital. In some cases, too, criticisms have been made that these improvements in productivity come at the expense of increasing dependence on the West or on international agencies. This is believed to be true, for instance, of the Green Revolution – one of the most publicised attempts at improving productivity and redressing inequalities in living standards in the developing world.

The Green Revolution showed that Malthus seriously underestimated how quickly knowledge – in agriculture, transportation and mechanisation – would transform food production. By the end of the twentieth century, world food supply was more than keeping up with population growth in most regions. The quest – breeding new seeds for enhanced agricultural productivity – was undertaken in the early post-Second World War years. These early stages, however, mostly involved narrowing the knowledge gap between what scientists already knew about plant genetics and the widespread ignorance on this score in developing countries. According to some authorities, this gap had been created and perpetuated deliberately by scientists and manufacturers in developed countries who thought they might not be able to make adequate profits out of new seeds which, once introduced, could be easily reproduced in developing countries. Gradually, however, higher-yielding varieties (HYV) of seed were made available, but they were initially only fully adopted by large landowning farmers with some education, access

Plate 7 *A foot-operated rice thresher being used in Yangshuo, Kwangsi-Chuang, China*

Photograph reproduced courtesy of Panos Pictures; © Sean Sprague

to credit facilities, and the capital to take risks with new seeds. Initially, at least, the Green Revolution resulted in some increases in inequalities: 'poor farmers, unable to borrow and lacking insurance or the savings to fall back on in the event of failure, could only watch and wait until the wealthier neighbours proved the value of the new seeds' (World Bank 1999: 5). The resulting time lag between the introduction of new seeds and their widespread use can be seen in the slow expansion of areas sown with new varieties.

The costs of these delays were significant. One study found that, for a farm family with 3.7 hectares, the average loss of potential increase over five years from slow adoption and inefficient use of HYV was nearly four times its annual farm income before the introduction of the new seeds. Eventually, however, the Green Revolution did boost the incomes of poor families and the landless. A survey in south India concluded that between 1973 and 1994 the average real income of small farmers increased by 90 per cent, and that of the landless – among the poorest in the farm community – by 125 per cent. Moreover, the poor benefited greatly from increased demand for their labour, because HYV demanded labour-intensive cultivating techniques. Calorie intakes for small farmers and the landless rose by 58–81 per cent, and protein intakes rose by 103–15 per cent (World Bank 1999: 5–6).

In assessing the impact of the Green Revolution, it is impossible to ignore the great benefits it has brought in improving productivity and the quality of farmers' lives in many parts of the developing world. The argument that it has encouraged dependency on developed countries and international agencies is certainly valid to some extent, especially in the early stages of its implementation, though in fact a great many non-profit organisations, private firms, banks, village moneylenders and land-rich farmers joined in the attempt to make the Green Revolution a success. But what experience of the Green Revolution *has* shown is that 'know-how' is only a part of what determines society's well-being. Information problems can very easily lead to market failures and impede efficiency and growth. According to one viewpoint, 'development thus entails the need for an institutional transformation that improves information and creates incentives for effort, innovation, saving, and investment and enables progressively complex changes that span increased distances and time' (World Bank 1999: 27).

The rational peasant

According to some writers, one of the main problems standing in the way of improving productivity through technological change is the resistance to change exhibited by peasants in many developing countries. It is argued either that the peasant is incurably lazy and unable to see the advantages of change, or that technological change is perceived to mean the destruction of social structures, something that peasants are unwilling to see happen.

However, it is now becoming more widely accepted that the peasant farmer is not necessarily acting illogically when he or she prefers to reject 'improvements' in agricultural practices. It is a fact that

> given the static nature of the peasant's environment, the uncertainties that surround them, the need to meet minimum survival levels of output, and the rigid social institutions into which they are locked, most peasants behave in an economically rational manner when confronted with alternative opportunities.
>
> (Todaro 1997: 318)

No farmers will have 'perfect knowledge', and the main motivation in a peasant's life may be the maximisation not of income, but rather of the family's chances of survival. Attitudes towards risks among small farmers may militate against apparently economically justified innovations. And many programmes to raise agricultural productivity among small famers have suffered because of failure to provide adequate insurance – including financial credit and physical 'buffer' stocks against the risks and uncertainties of crop shortfalls.

Central to an understanding of this issue is what has been called 'income smoothing'. When householders cannot, because of poverty, smooth their consumption through the normal coping mechanisms of loans, asset sales, grain storage and transfers from family and neighbours, they smooth their income instead by avoiding risky, but on average more profitable, opportunities. For example, one study in central India found that small farmers – a group with limited capacity to smooth consumption – planted only about 9 per cent of their land with relatively risky, high-yielding varieties, whereas large farmers, with better access to coping mechanisms, planted about 36 per cent with high-yielding seeds (World Bank 1999: 123).

We may conclude that peasant farmers do act rationally and that they are ready to respond to economic incentives and opportunities where these are perceived to be in their real interests. Efforts to minimise risk and to remove commercial and institutional obstacles to small-farmer innovation are therefore essential requirements of agricultural and rural development planning.

Women's roles in rural-agricultural development

In many parts of the developing world women undertake many of the tasks in agriculture and are especially responsible for basic food supplies. Women provide 60–80 per cent of agricultural labour in Africa and Asia, and about 40 per cent in Latin America. As Boserup points out, women normally work longer hours and perform many roles outside agriculture: reproduction, labour for cash crops, food crops for the home, raising and marketing food and livestock, engaging in many cottage industries such as pot-making and cloth-making, as well as all household duties such as preparing and cooking food, and collecting firewood and water (Todaro 1997: 314).

However, the role of women tends to be ignored or marginalised in planning policies in developing countries and, as the work of Chambers referred to at the beginning of this chapter shows, urban bias works against the support of women. But the active participation of women is critical to agricultural prosperity, and policy design should ensure that women benefit equally from development efforts.

Environmental issues

Reference has already been made to the climatic hazards affecting agriculture, and the problems of environmental pollution will be discussed in the next chapter. Here, however, brief mention must be made of one issue – deforestation – that has attracted a good deal of attention in the literature and in the public imagination.

The problem of deforestation can occur at the very local level, where a village population dependent on firewood for its domestic fuel eventually becomes aware of serious shortages. But most popular attention has been directed at the large-scale destruction of forests, notably in the Amazon Basin, and in the effects on global climate that massive deforestation might have.

This problem has been examined in Chile. Chile has a reputation as Latin America's economic success story of the 1990s. But this impressive record of sustained economic development, coupled with improvements in social justice, has incurred significant environmental costs that raise doubts about the ecological sustainability of the Chilean model of resource-led development. For Chile's economic growth has been led by exports of natural resources, including the Valdivian rain forest, in which three out of the four major forest types are already in danger of destruction. The native ecosystem has been badly affected. Forestry exports from Chile tripled between 1987 and 1994, and it is possible that forestry exports may soon outstrip copper as Chile's main export-earner. This resource-based, export-led strategy is dangerous: 'the current path of forestry in Chile is unsustainable and threatens the existence of its forest ecosystems' (Clapp 1998: 30). A new forest law is urgently needed.

Another, rather different problem of forest exploitation is found in Cambodia. Here the main problem is illegal logging. Foreign donors have now made forestry reform in Cambodia a condition of their offer of aid. For if felling continues at the present rate, reckons the World Bank, the last tree will be cut down in about four years' time. But here, as elsewhere, the forests are seen very differently by politicians, soldiers and businessmen on the one hand and by environmentalists on the other. In Cambodia forestry represents 43 per cent of foreign trade by value. The main money-spinner is the high-quality teak and other hardwoods sent to Europe and America. Demand from South Korea and Japan is expected to rise, and a vast new potential market is China. In July 1999 China imposed a domestic logging ban after the Yangtze River floods were blamed on 'rampant' logging that had caused soil erosion and prevented water retention. The lack of wood has become serious, and imports into Shanghai from Malaysia and Cambodia have increased rapidly. This is attractive to illegal loggers in Cambodia, Myanmar and Laos. According to the World Bank, Cambodia lost $60 million in revenue from illegal logging in 1997: the money went into the pockets of corrupt individuals and the army, rather than into the state coffers.

If illegal logging is not stopped, say environmentalists, large parts of South-East Asia may suffer from flooding, erosion and other afflictions that have hit heavily deforested areas in countries such as China and the Philippines. Reform is possible if foresters learn to manage trees as a renewable resource and can meet international standards. Their progress in doing so can be policed by satellites and timber-tracing. Aid can also be made conditional on reforms.

Plate 8 *New Guinea hut being built from local forest materials*
Photograph: B.W. Hodder

Rural development

According to one common perspective on development, the core problems of widespread poverty, growing inequality, rapid population growth and rising unemployment all find their origins in the stagnation and often retrogression of economic life in rural areas. In the 1960s and 1970s there was an attempt to bring about improvements in rural life by a policy of integrated rural development. Integrated rural development programmes required (a) accelerated output growth by small farmers, (b) rising domestic demand, and (c) diversified, non-agricultural labour-intensive rural developments which support and are supported by farming communities. It has had only limited success, however, and there has been a tendency to move away from the government-sponsored integrated rural development schemes towards similar operations run by, for example, non-governmental organisations (NGOs). The aim of NGOs is not simply to increase production but also to improve the quality of life. It is appreciated that rural development is not just a matter of raising productivity; it also requires off-farm employment creation and education, health and social services (Todaro 1997: 296). The role of NGOs has, indeed, increased markedly over the past few years. Now, however, NGOs are coming under criticism as their operations become larger, less voluntary, more bureaucratic and more associated with governments.

Furthermore, as Sundberg (1998) has pointed out for the Maya Biosphere Reserve in Guatemala, the misdirected projects and landscapes there reflect the goals of NGOs and not the people's needs or aspirations. And in another case, Fyvie and Ager (1999) have examined NGOs and innovation in the Gambia; they conclude that for development assistance work in that country, innovation – often claimed to be a key attribute of NGOs working in the field of development – has been far less effective than it should have been.

Agriculture or industry?

Before discussing briefly the mutual interdependence of the rural-agricultural and urban-industrial sectors, it is important to reiterate the point made earlier that there has been a tendency to underplay the importance of agriculture in development policy. Part of the reason for this has been the recognition that the share of agriculture in national income generally declines as per capita incomes rise. And the reasons for this relative decline are, first, that people tend to spend less on food as a percentage of their incomes as these incomes rise. Second, as farmers increase the productivity of their land and labour, the share of a country's resources required to grow food for the rest of the population decreases.

The viewpoint accepted in this chapter – that rural agricultural development involves a change from simple peasant agriculture to modern commercial or industrialised agriculture – means that the point of transition from agriculture to industry is not always easy to identify. It is true that governments frequently have to make choices about whether to emphasise agriculture or industry in their development planning, but the links between the two sectors are so strong that they may be regarded as mutually interdependent. Each plays a strategic role in the development of the other for the following reasons:

1 Agriculture depends for its development upon the provision of manufactured goods, whether for the direct transformation of agriculture (farm tools or machines, tractors, fertilisers, irrigation pumps and the like) or for consumer goods such as radios and bicycles which are in demand as incomes rise. In other words, the agricultural sector provides a major market for the urban-industrial sector and thus can stimulate urban growth and manufacturing industry within a country.

2 As agricultural development through technological change proceeds, so does labour become less significant a factor. Labour and capital are thereby released and move into the urban-industrial sector of an economy. This is an important point, illustrating as it does the truth that industrialisation (i.e. technological transformation) of agriculture must invariably lead to a decline in the farming population, part of which is thereby released from agriculture and moves into the industrial sector (Table 4.2). This has been as true of Western Europe as it is of developing countries today.

3 Agriculture can provide the raw materials for industry – tobacco, cotton and sisal, for example – which otherwise would have to be imported from abroad.

4 Where agriculture produces export crops, such as cocoa, coffee or palm oil, the foreign exchange can be used for importing items used for industrial development, such as petroleum and chemicals, which cannot be produced locally. A problem here, however, is that while increasing cash crops for export increases foreign exchange earnings, this might be at the expense of producing foodstuffs for the domestic market (Box 4.2).

5 The relationship between prices and taxation policy as these affect the rural and urban sectors respectively is important and can affect motivation in quite critical ways. Thus high prices for agricultural crops, including food, are welcome to farmers, who will produce and sell more; but if this means high prices in urban areas, then poverty and unrest can develop in the urban industrial centres. On the other hand, low food prices in the urban industrial sector can result in farmers producing less food. As for taxation, tax in the agricultural sector may be necessary to raise revenues to finance public expenditure, but higher taxes may also act as a disincentive to farmers.

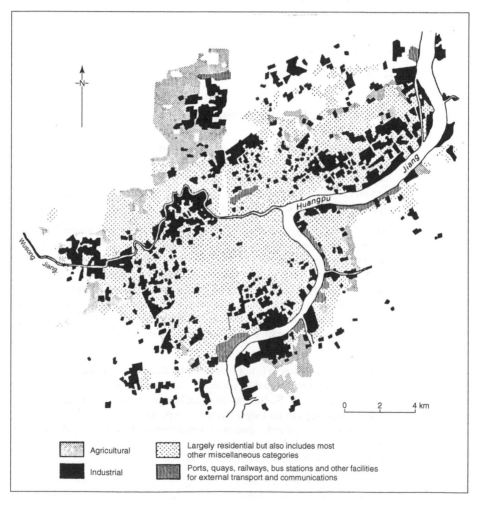

Figure 4.1 *Land use in and around Shanghai City*
Source: Hodder (1989)

Table 4.2 *Changing percentage of the economically active population employed in agriculture*

Country	1980	1990	2000
Thailand	71	63	58
Indonesia	58	52	48
Philippines	53	44	39
Malaysia	40	25	18

Source: FAO; *The Economist*, 3 April 1999 (forecast for 2000)

6 Agriculture performs a critical role in providing food for the increasing urban populations involved in industrialisation. The example that follows makes this point, as do a number of other examples referred to in the previous three chapters. It also sets the scene for our study of urban-industrial development in Chapter 5.

Box 4.2

Food production in Senegal

Peasants in many parts of Senegal are reducing their groundnut plantings and planting more of the traditional food crops: millet and sorghum. The reason is simple: they are unable to earn enough from the sale of groundnuts to buy the food they need for themselves, so it is safer to grow their own food at the expense of the major cash crop of Senegal.

This, of course, has the effect of reducing exports, so the government has less foreign exchange with which to import food for the urban population or to help farmers increase their productivity. It is a vicious circle with no obvious way out. According to one authority, groundnut producers earned more for their daily labour in 1913 than in 1984, and things have got worse, if anything, since then. But with a population of more than a million dependent on it, there is no way in which government can abandon the groundnut economy; they will simply have to keep propping it up with costly price subsidies, just to prevent a total collapse.

Yet even with subsidies, it is not economic for farmers to use the quantity of chemical fertiliser that would be necessary to protect the soil, or to leave land fallow to restore itself. So large areas of arable land are being 'mined' of their nutrients until they can no longer sustain a crop, and then abandoned. According to the Food and Agriculture Organization (FAO), if soil degradation in this part of the Sahel is not controlled, the region will soon be able to feed only half the number of people it used to support.

The government's declared aim of reaching 80 per cent food self-sufficiency by the year 2000 would require the rural population to produce almost twice its own needs, in order to feed the towns. In fact current production trends are going in the wrong direction – and an official campaign which has been trying for several years to persuade urban consumers back to traditional cereals, in order to reduce imports of wheat and rice, has had very limited success.

Rice and wheatflour are the city's convenience foods. Housewives tend to reject millet and sorghum because they take too long to prepare and cook. Even in small provincial towns the French baguette from a local bakery has become the symbol of modern living. Some bread may be made from Food Aid wheat, but an increasing share of imports has to be paid for.

Source: Sharp (1994: 28)

Peri-urban vegetable gardening in China

Chinese peri-urban vegetable gardening is often cited as perhaps the most intensive and, within its technological parameters, the most efficient and sophisticated traditional system of farming in the world. With the rapidly increasing urbanisation in China, the problem is how to raise the production of fresh vegetables for the urban markets – for fresh vegetables are an essential ingredient of any Chinese diet. Distance from the urban centre determines not only whether vegetables are grown, but also what type of vegetables are grown for the nearby urban market. The central issue here, however, is whether even greater yields and productivity could be achieved largely on the basis of the existing indigenous technology – upgrading it with the application of modern inputs, such as chemical fertilisers and higher-yielding, more pest-resistant, drought-resistant and cold-tolerant plant varieties – or on the basis of modernising or 'industrialising' vegetable farming by establishing larger, specialised units of production where the advantages of scale, large-scale mechanisation and irrigation control can be added to other inputs.

Arguments for perpetuating yet at the same time upgrading the indigenous small-scale and essentially labour-intensive technology of traditional Chinese vegetable farmers are numerous. Apart from a natural reluctance to lose this unique and traditional form of farming, with all its cultural associations; and apart, too, from the fact that this form of farming has been emulated in and successfully transferred to many other parts of South-East Asia, Africa and elsewhere, other, more cogent arguments are frequently advanced. It is pointed out that the Chinese vegetable farmers have already shown that they can adapt quickly and successfully to the growth of urban demand for fresh vegetables, and can continue to do so, given proper control of price mechanisms and

Plate 9 *Peri-urban vegetable cultivation in Lome, Togo, West Africa. This form of agriculture was introduced here, and elsewhere in West Africa, by the Chinese*

Photograph: B.W. Hodder

adequate marketing and distribution networks. It is also argued that Chinese traditional techniques are ecologically sound, are less destructive of the environment, and are less wasteful in the use of fossil fuels.

Those who support a change to a more modernised or 'industrialised' form of farming, based on larger, more coherent and specialised land units and the application of the larger and more expensive forms of modern inputs, argue that the traditional Chinese agricultural technology has already reached its peak of efficiency and is approaching a point of diminishing returns. With larger, capitalised units, the mechanisation of almost all farming operations is possible, and both physically – the mechanisation of farming on small plots, with a close network of irrigation canals, trellising and multi-cropping is commonly very difficult – and economically worthwhile. Then, too, the value of the application of scientific farming methods, the use of chemical fertilisers, pesticides and the introduction of new strains of crops has already been proved beyond doubt. Finally, modern industrialised farming is essentially labour-saving, and there is already considerable evidence of labour shortages in peri-urban vegetable farming in and around Chinese urban centres.

These labour shortages reflect wide differentials in the demand for labour between rural and urban activities, discouraging farmers from working their vegetable farms and encouraging them to move into more rewarding urban, non-farm occupations. But another important factor impelling farmers out of vegetable farming is that it is extremely hard work. Moreover – and this illustrates a point made in the previous chapter – in some parts of China vegetable farming carries low prestige compared with, for instance, the cultivation of rice, even though the returns from rice may be lower. As several writers have noted, vegetable farmers have lower status than rice farmers, even though their cash returns are higher, because vegetable farming is known to be hard, requiring continuous, painstaking labour. The traditional Chinese intensive 'garden' form of farming did not arise out of ignorance, or from any belief in the work ethic. Like similar systems elsewhere, such labour-intensive cultivation arose out of necessity. Thus whatever the ecological or other merits of this traditional form of agriculture, it is not a type of farming likely to be continued for its own sake. Whatever Western observers may say about it, Chinese farmers finds the work hard, tedious and continuous; they do it only because they have to in order to live; and if they can make an adequate living in some less exhausting occupation, then they have shown themselves only too willing to change.

Shortage of labour for vegetable cultivation around urban areas in China is becoming severe today and may well become the major factor directing change in the nature and location of vegetable farming for urban markets in China. If this is so, then the days of peri-urban vegetable farming in China would seem to be numbered. Specialised market gardening or 'trucking' zones are likely to develop in exactly the same way, and for exactly the same reasons, that such zones have emerged elsewhere (Hodder 1987).

Summary

This chapter notes the tendency to 'urban bias', in which increasingly less attention is given to rural-agricultural development. The main aim now seems to be to effect the transition from simple peasant agriculture to commercial or 'industrialised' agriculture. Following a summary of the major types of agriculture in the developing world, a number of problems in effecting the transition to commercial agriculture are highlighted, including land reform, changing technology, the Green Revolution and environmental concerns. The debate over 'agriculture or industry' is referred to and related to the changes now taking place in agricultural policy. The case of peri-urban vegetable farming in China is presented as an example of the kinds of changes and problems facing most agricultural systems oriented towards urban markets.

Questions for discussion

1 What are the main differences between Latin America and the other two major sectors of the developing world in terms of agricultural development?
2 Identify and illustrate what is meant by 'urban bias'.
3 To what extent do you agree that agricultural development is only partly a matter of technological change?

Further reading

Adams, W. (1993) 'Indigenous use of wetlands and sustainable development in West Africa' *Geographical Journal* 159: 209–18.

Bebbington, A. (1993) 'Governments, NGOs and agricultural development: perspectives on changing inter-organizational relationships' *Journal of Development Studies* 2: 199–219.

Grigg, D. (1992) 'World agricultural production and productivity in the late 1980s' *Geography* 77: 97–108.

Hodder, R. (1987) 'Some dynamics of peri-urban vegetable farming in China' MPhil. thesis (unpublished), University of Hong Kong.

Morgan, W.B. and Solarz, J.A. (1994) 'Agricultural crisis in sub-Saharan Africa: development constraints and policy problems' *Geographical Journal* 160: 57–73.

Powell, S.G. (1992) *Agricultural Reform in China: From Communes to Commodity Economy* Manchester: Manchester University Press.

Rigg, J. (1989) 'The Green Revolution and equity: who adopts the new rice varieties and why' *Geography* 74: 144–50.

Todaro, M. (1997) *Economic Development* London: Longman (chapter 9).

5 ▶ Urban-industrial development

- ● Urbanisation: levels and dynamics
- ● A Third World city?
- ● Urbanisation and industrialisation
- ● The informal sector
- ● Problems and policies
- ● Urbanisation and development

This chapter focuses on a number of issues concerning urbanisation, industrialisation and development, although, as indicated in Chapter 4 and in Box 5.1, agriculture remains an important occupation in many urban centres in the developing world (see also Figure 4.1). The first section presents a summary, together with statistical evidence, of some of the main points about this topic normally discussed at greater length, but available elsewhere. As with many aspects of 'urban geography', these have been presented in some detail in another volume in the same series as this book: *Urban Geography*, by Tim Hall. In his book, Hall uses the classification of different types of cities made by Savage and Warde (1993): Third World cities, cities in socialist countries, global (world) cities, older (former industrial) cities, and new industrial cities. He focuses primarily on the urban geography of the last three of these types. In

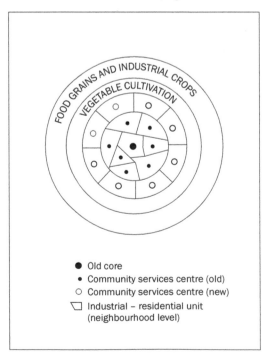

● Old core
● Community services centre (old)
○ Community services centre (new)
⊐ Industrial – residential unit
  (neighbourhood level)

Figure 5.1 *Land use and functional zones in and around a Chinese city*
Source: Hodder (1989)

Box 5.1

Agriculture as an urban occupation

A common feature of many cities in developing countries is the often surprising role agriculture still plays in the urban economy. The cases of Shanghai in China and Ibadan in Nigeria are often mentioned in this regard (Yeung 1998; Mabogunje 1989). Another example is Davao City in the Philippines (Figure 5.2), The city of Davao lies on the south-eastern coast of the island of Mindanao in the south of the Philippine archipelago. The city is the second largest in the Philippines after Manila. Large rural areas are encompassed within Davao's administrative jurisdiction. Most of the urban and urbanising parts of the city, however, are on the coastal plains, which form a narrow band averaging 5 km in width and extending 36 km from Lasang to Binugao. The city proper occupies a thin 25-km slice of this coastal strip, and although it covers only 3.6 per cent of the total area of Davao City, it contains about 85 per cent of the total population living in urban and urbanising areas, or about 60 per cent of the total population of Davao City. The scattered distribution of the squatter settlements is clearly evident from Figure 5.2.

Agriculture still forms the mainstay of Davao's economy, for although only about a quarter of the labour force is employed in the primary sector, the secondary and tertiary sectors are both highly dependent on the primary sector. Moreover, agricultural goods are the most important source of foreign exchange. The port of Davao City – the largest port outside Manila – handles around 70 per cent of all exports from the southern region, and about two-thirds of this consists of agricultural goods, mainly fresh bananas, coconut products, coffee, pineapples, fish and crude palm oil.

Source: Hodder (1991)

---

these pages we consider especially Third World cities, the largest of which are of course also 'world' cities, but the main emphasis here is on the problems, processes and policies of urbanisation and industrialisation and how these link into and affect the nature and extent of development, both locally and globally. This is central to any study of development geography: for many writers development is almost synonymous with urbanisation and its associated industrialisation.

Urbanisation: levels and dynamics

All data on the levels and dynamics of urbanisation (Table 5.1) need to be accepted with some caution for two main reasons. First, the reliability

Table 5.1 *Urban population: changes by region*

Region	Urban pop., 1985 (million)	Urban pop. growth rate, 1970–85 (%)	Projected urban pop., 2000 (million)	Projected urban pop. growth rate, 1985–2000 (%)
East Asia	46	4.4	68	2.6
S and S-E Asia	377	4.1	694	4.2
West Asia	63	4.6	109	3.7
Latin America	279	3.6	417	2.7
Africa	174	5.0	361	5.0
Pacific	1	4.2	2	4.7
China	219	1.8	322	2.6
Total	1 159	3.7	1 972	3.6

Source: Devas and Rakodi (1993); Elliott (1999)

of census data varies very significantly from country. Second, the definition of 'urban' varies, sometimes quite dramatically, from country to country. On the global scale all one can do is to work with the data produced in the annual United Nations and World Bank publications. These data provide evidence for some of the main summary points made below about the levels and dynamics of urbanisation.

Levels of urbanisation

- Of the major regions in the developing world, Latin America, with an urban percentage of 73 per cent, is easily the most urbanised (and industrialised), which goes some way to explaining why much of the literature on urbanisation in the developing world is based on work in Latin America.
- Of the remaining regions, urban percentages range from the lowest (South Asia, at 25 per cent urban) to the Middle East/North African region, at 55 per cent urban.
- Well over half of the world's cities with a population of more than a million are now in the developing world. Of the world's twenty-five largest cities, twenty are in developing countries – four of them (Buenos Aires, Rio de Janeiro, São Paulo and Mexico City) in Latin America.
- There is a high degree of primacy (a disproportionate population in the largest city) in most developing countries.

Dynamics of urbanisation

- It is estimated that at present (the year 2000), 66 per cent of the world's urban population live in developing countries and that by 2025 the percentage will have risen to 77 per cent.
- In terms of regions, by 2025 Africa's percentage of the world's urban population will have risen from 10 per cent to 17 per cent; Latin America's share will have dropped slightly from 13 per cent to 11 per cent; and Asia's share will have risen from 40 per cent to 49 per cent.
- Rates of urban growth are much higher (up to 7 per cent per annum) in the cities of the developing world than in the developed world, where rates of increase are sometimes below 1 per cent per annum.

A Third World city?

What are commonly regarded as the characteristics of Third World cities can be simply summarised as follows:

- They have a significant informal sector (defined below).
- They have a heterogeneous population, including a significant immigrant population.
- Many have an essentially pre-industrial economic base, including agriculture (Box 5.1).
- They are largely coastal in location (reflecting colonial dependency).
- Many have colonial accretions.
- It is common for them to have illegal squatter settlements and shanty towns (with a poor fabric).
- They have high levels of unemployment and poverty.
- They show little development of a central business district (CBD).

However, the answer to the question 'Is there a Third World city?' depends ultimately on whether the characteristics of these cities, summarised above, are sufficiently distinctive from those of cities in the developed world. To take an example, are Singapore or Buenos Aires distinctively Third World cities compared with, say, Sydney in Australia? One's answer also depends on how far cultural characteristics can be separated out from other elements and what part colonial contact has played.

It may be that a positive answer to the question 'Is there a Third World City?' is possible only within the context of a specific region, such as

Plate 10 *Housing development in Singapore*

Photograph: B.W. Hodder

South-East Asia. There has as yet been no successful attempt to construct a more generalised model; indeed, a number of writers have suggested that the Latin American city has characteristics that are different from those of other parts of the developing world (Potter and Lloyd-Evans 1998).

Urbanisation and industrialisation

It is a generally accepted principle that the benefits of economic development are better absorbed and distributed through urban centres than by any other means. It is also commonly argued that urbanisation and industrialisation are virtually synonymous processes. Cities and city regions are identified as the catalyst of a process designed to engage the whole national space:

> Economic growth tends to occur in the matrix of urban regions. It is through this matrix that the evolving space economy is organised. The location decisions of most firms, including those in agriculture, are made with reference to cities or urban regions.
>
> (Friedmann 1966: 28–9)

As noted in Chapter 4, the share of the industrial sector tends to increase at the expense of the agricultural sector as an economy develops. While this has been generally true throughout the developing world, the experience of industrialisation has been very uneven. Latin America has advanced furthest in its industrialisation, but the most rapid and successful industrialisation has occurred recently in East and South-East Asia, though in 1998–9 the economies of these regions have experienced a downturn.

In Latin American and many other developing countries, initially the strategy for industrialisation was import substitution industrialisation (ISI). This strategy was based on the domestic production of goods that were previously imported, and involved raising tariffs on imports so as to protect the domestic market for domestic firms or the affiliates of multinational corporations. The strategy involved three stages: the manufacture of consumer goods, the production of investment goods, and finally the production of capital goods. But in practice few countries ever got beyond the first stage, and economic failure led to unrest and, in Brazil, a coup. Why the ISI strategy failed is a matter of opinion. Some argue that there was too much government intervention, while others have a structural explanation: a shortage of foreign exchange, dependence on inappropriate foreign technology, and an inability to deal effectively with bottlenecks.

A quite different strategy of industrialisation has had rather more success and is believed to have been largely responsible for the success – however temporary – of the so-called tiger economies of the Far East – Taiwan, South Korea, Hong Kong and Singapore. In this strategy – export oriented industrialisation (EOI) – emphasis is laid on outward-oriented trade policies, the use of labour-intensive technologies in manufacturing, reliance on the market for the allocation of resources, and minimal government intervention. This strategy was also taken up by the newly industrialising economies (NIEs) of Malaysia, Thailand, Indonesia and China.

Opinions on the merits of EOI vary quite widely. Some writers go so far as to argue that the export-oriented industrialisation practised in Taiwan and South Korea in the 1960s has given us a model of urban-industrial success. The sequence of stages followed there was:

- labour-intensive economic activity;
- capital-intensive, raw-material intensive activity;
- capital-intensive, machinery-intensive economic activity; and
- knowledge-intensive activity.

Some argue that this could be the way out of poverty for all Third World countries. On the other hand, the East Asian crises of the past two years or so have cast doubts on EOI as a necessarily more 'successful' strategy than ISI. And it has been argued that the EOI strategy, based on cheap labour, has been characterised by a good deal of social unrest and political repression.

The informal sector

It has become common to divide up the urban economy into formal and informal sectors, and, in spite of the many criticisms levelled at this distinction, it is still a useful concept in trying to understand the nature and problems of cities in developing countries (Table 5.2). The informal sector is linked both with the rural sector (from where most members of the informal sector come) and with the urban formal sector, which depends on the informal sector for many services and certain elements of labour. Typically the informal sector 'is characterised by a large number of small-scale production and service activities that are individually

Table 5.2 *Characteristics of the urban formal and informal sectors*

The formal sector	The informal sector
Difficult entry	Easy entry
Reliance on overseas resources	Reliance on indigenous resources
Corporate ownership	Family ownership of enterprises
Large-scale operation	Small-scale operation
Capital-intensive production	Labour-intensive production
Use of imported technology	Use of traditional technology
Formally acquired skills and qualifications	Skills acquired outside formal system
Protected markets (tariffs and licences)	Unregulated and competitive markets

Source: Potter and Lloyd-Evans (1998)

owned or family owned and use labour-intensive and simple technology'
(Todaro 1997: 269). Their activities range from hawking and street
vending to letter-writing, haircutting and small artisan activities. Without
capital, legal status, or official work of any kind, members of the
informal sector usually have no proper accommodation and live on
pavements or in slums and shanty towns. They work long hours and have
very low standards of living and security.

Until recently the informal sector was given little attention except as a
social problem and as a transitional phenomenon that would eventually
be absorbed into the formal sector in a city. Now, however, it is
recognised that the informal sector has a great deal to contribute to the
economy of Third World cities. The estimated share of the urban labour
force in the informal sector ranges from under 20 per cent in Colombo
(Sri Lanka) to over 60 per cent in Kumasi (Ghana) and urban areas in
Pakistan and Peru. The informal sector plays an important part in
providing income opportunities for the poor, and support is growing for
the view that the informal sector should now be recognised as a
permanent feature of the urban economy and should be promoted as a
major source of employment and income for the urban labour force.
According to some estimates, the informal sector is already providing an
average of 50 per cent of the urban labour force and generates almost
one-third of urban income.

Some writers express the fear that encouraging the informal sector could
aggravate urban unemployment by attracting too much labour from rural

areas. Pollution could also be increased, as could congestion and a whole range of problems such as housing and the provision of other public services (Box 5.2). But the above comments can all tend to perpetuate the notion that the informal sector is a discrete category of marginalised activity. More accurately, it has been represented as part of the informal–formal sector continuum, ranging from the subsistence sector, through small-scale producers and retailers, to the third category of 'petty capitalists', and to full capitalist production (Potter and Lloyd-Evans 1998: 178). And Freeman (1996) notes that the Doi Moi (Renovation) boom in Vietnam – a sudden reversal of Communist Party policy in 1986 – led to a boom in Ho Chi Minh City, and that the informal sector, characterised by numerous petty entrepreneurs, was the agent leading to Doi Moi, which then began to lay the foundations for successful liberalisation of national economic policies.

Box 5.2

Senegal's New Industrial Policy and the informal sector

The Senegal government's New Industrial Policy, launched in 1984 under the structural adjustment regime, aimed to attract new industries to the country, to privatise all activities that could be detached from the public sector, and to lower customs barriers, exposing local producers to foreign competition.

This was a dramatic about-turn from previous industrial policies, which were based on the protection of local industry and socialist principles of state management. Neither the protection nor the principles had really worked: labour productivity had been declining in most industrial sectors since the mid-1970s, but, as in agriculture, no one was prepared for the sudden change. A further shrinkage in jobs has thus meant a rapid expansion of the informal sector – especially in Dakar and the other main towns – and a rise in urban unemployment.

There were 143 000 people out of work in Dakar alone by 1991 – almost 25 per cent of the active population, according to a recent survey. Nationally, unemployment levels had already reached 20 per cent by 1988. This would be bad enough if the country's workforce were not increasing; in fact, it is estimated that 100 000 young people are entering the job market every year. With recruitment to the public sector virtually frozen and formal jobs in the private sector declining, the chances of finding work are now better for those with less education. Despite the remarkable Senegalese capacity for getting by somehow, this state of affairs poses a serious threat to peace and social stability.

'The informal sector' is a dry, academic term to describe the hive of makeshift livelihoods – market traders, artists, hairdressers, transporters, dressmakers, builders and traders of all kinds – which make life possible for most of the time. Partly because of the name, the informal sector is usually regarded as subsidiary to the formal economy of 'proper' businesses and jobs. Actually it is the other way round. There were about 600 000 people with jobs in the informal sector in Senegal urban areas in the late 1980s, compared with 200 000 in the formal economy. Since then, job losses have pushed the number close to 1 million.

As businesses and employment have shrunk, so have the government's tax receipts. To make up the deficit, it can choose between taxing the rest more heavily (which could be counter-productive), or closing the system's many loopholes (which would offend many of the government's supporters), or finding new sources of revenue. The World Bank suggests ways of finding ways to tax the informal sector. But this could stifle its vitality, or force it further underground.

Source: Sharp (1994: 52)

Regarded in this light, the informal sector clearly has much to offer, and would have still more if it were given access to the advantages enjoyed by the formal sector, such as credit, foreign exchange and tax concessions. The informal sector is capital-saving; investment in human capital is cheap in the informal sector; and it provides unskilled and semi-skilled labour, used to using appropriate technology. In considerations of policy towards the informal sector, the place of women in the sector is also important, especially where women make up the majority of those migrating to the city and finishing up in the informal sector. They are usually disadvantaged compared with their male counterparts and work for low wages, with no security or employee benefits of any kind. Furthermore, they are rarely the recipients of any improvements directed at the informal sector in terms of, for instance, micro-credit facilities. The vast majority of institutional credit is channelled through formal-sector agencies, and 'women generally find themselves ineligible for small loans' (Todaro 1997: 273).

Problems and policies

Migration to urban centres

The high rate of growth in the cities of developing countries is due not only to relatively high rates of natural increase but also, and in many cases more importantly, to rural–urban migration. Very often, natural increase and in-migration contribute in broadly equal proportions to the total growth of urban populations. Thus for Brazil in the 1970s in-migration accounted for 2.2 per cent per annum out of a total urban growth rate of 4.4 per cent per year (Potter and Lloyd-Evans 1998: 13). It is difficult to find reliable comparative data on this point, but, using 1980 data, one author estimates that the share of urban growth due to migration varies from 35 per cent in Argentina to 64 per cent in Nigeria and Tanzania.

The causes of migration to the city have been examined by a number of authors, and it is recognised that an understanding of this phenomenon is central to any analysis of employment problems in the urban areas of developing countries:

> if migration is a key determinant of the urban labour supply, the migration process must be understood before the nature and causes of urban unemployment can be properly understood. Government policies to ameliorate the urban unemployment problem must be based, in the first instance, on knowledge of who comes to town and why.
>
> (Todaro 1997: 278)

In his well-known rural–urban migration model, Todaro assumes that migration is primarily an economic phenomenon, which for the individual migrant can be a quite rational decision despite the existence of urban unemployment. He postulates that 'migration proceeds in response to urban–rural differences in *expected income* rather than actual earnings'. Migrants 'consider the various labour market opportunities available to them in the rural and urban sectors and choose the one that maximises their expected gain from migration' (ibid.: 280).

Some writers, however, believe that such economic models reduce migration to a 'straightfoward choice between a traditional, backward, rural way of life and a modern, industrialised lifestyle in the city'. But in fact,

> urban values and styles of living long since penetrated even the remotest regions; a process which is reinforced by return migration. In parallel, traditional, rural values have moved into the city with the migrants . . . in many ways it is the blurring of distinctions which has served to encourage migration because there is not now such a drastic change of lifestyle involved.
>
> (Drakakis-Smith 1987: 33)

Another writer describes rural–urban population transfers as a 'circular, interdependent, progressively complex and self-modifying system in which the effects of changes in one part of the system can be traced through the whole system' (Mabogunje 1980: 30). The perception of opportunities to earn a higher wage, or to achieve a more desirable lifestyle in urban areas, is likely to attract migrants, who will themselves become sources of information about urban opportunities for their families and neighbours living in rural areas.

Whatever the causes or characteristics of rural–urban migration, the evidence now available suggests that, while still important, migration is declining as a factor in rapid urban growth, especially in Africa, Asia and the Middle East, where much migration is often circular and a largely male phenomenon. In Latin America and the Caribbean, however, female migration is more common.

The desire for education can be a powerful motivating force for migrants, but in some cases migration to the cities is forced by military or political action. In Angola in 1999, for instance, UNITA (União Nacional para a Independencia Total de Angola) rebels bombarded the already devastated cities of the interior. Some 600,000 members of farming families have been forced to abandon their homes and their crops by the fighting between the rebels and government forces in the countryside. These families are driven into the cities, where they are dependent largely on international food aid. The rebel leader apparently wants the population to be forced into the government-controlled cities, thus swamping the virtually non-existent public services and provoking opposition to the government.

In Asia, the Middle East and Africa, most migrants tend to be men, their womenfolk remaining on the farms, maintaining the cultivation of subsistence crops, and the security of the villages to which the men can return. In Latin America and the Caribbean, however, women are more numerous, owing to the historical process of male and female proletarianisation: women withdraw their labour from agricultural labour and there is the stereotyping of domestic work as women's work. In all cases links are often maintained with their home areas by migrants in the cities, who send remittances back home or who band together in regional associations or tribal language associations, and may raise funds for development projects in their areas of origin.

Housing: spontaneous self-help units

In discussing the housing of the poor in cities in developing countries it is usual to distinguish between shanty towns and squatter settlements. The former are defined by the poor materials used to construct some kind of shelter, whereas squatter settlements are housing built illegally on ground the squatters do not own. However, as Potter and Lloyd have noted, the various labels to describe different types are misleading and ignore the dynamics which take place and the enormous diversity with regard to formation, building materials, physical character and their inhabitants (Potter and Lloyd-Evans 1998: 138–41).

While there is general agreement that shanty towns are unacceptable places for people to live, there is some disagreement over the role squatter settlements can play in dealing with the housing problem in Third World cities. Are squatter settlements the problem or the solution? They are now often viewed as 'slums of hope' – residential areas where populations begin to adapt to city life (Mangin 1967; Lloyd 1979) – staging posts on the way to being integrated fully into city life. In Davao City in the Philippines, sixty-nine squatter settlements, comprising a total of 18 000 units, were identified in the city proper alone (Figure 5.2). The combined population of these settlements was estimated to be over 100 000 – or about 25 per cent of the total population of the urban and urbanising areas of Davao City.

Planning to deal with the 'squatter problem' in the city has not been very clear. In Davao City the Community Mortgage Programme was designed to assist squatters to buy the land they occupy and to construct decent, affordable houses. From the point of view of the squatters, however, the logic of the government's action is perverse. While there are admittedly many flimsy, makeshift houses, there are also plenty of good-quality wooden or concrete houses. Moreover, in some of the large squatter settlements there are also good-quality commercial buildings. Indeed, it is often hard to tell where a squatter settlement, as officially designated, begins or ends. Many of the squatters are young, middle-class educated professionals – including doctors, dentists, engineers, lawyers, middle-level managers and administrators, and business people – as well as a host of small entrepreneurs ranging from storekeepers to taxi-drivers. Squatters are the *de facto* owners of the land and many of the houses were constructed cheaply, quickly and without the need to follow bureaucratic precedures. The residents' opportunities for material and personal success are seen to lie with the individual, not with the

Plate 11 Squatter settlement in Davao, the Philippines

Photograph: R.N.W. Hodder

Plate 12 *Squatter settlement in Davao, the Philippines*
Photograph: R.N.W. Hodder

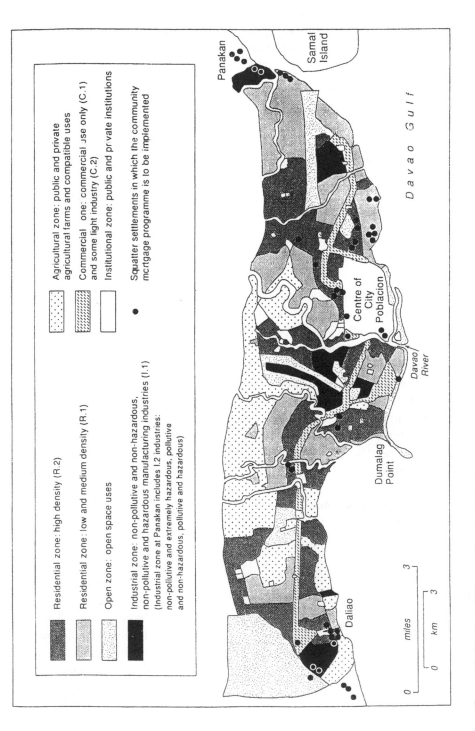

Figure 5.2 *Land use in Davao City, the Philippines*
Source: Hodder (1991: 105)

Key (as shown in legend):

Residential zone: high density (R.2)

Residential zone: low and medium density (R.1)

Open zone: open space uses

Industrial zone: non-pollutive and non-hazardous, non-pollutive and hazardous manufacturing industries (I.1) (Industrial zone at Panakan includes I.2 industries: non-pollutive and extremely hazardous, pollutive and non-hazardous, pollutive and hazardous)

Agricultural zone: public and private agricultural farms and compatible uses

Commercial one: commercial use only (C.1) and some light industry (C.2)

Institutional zone: public and private institutions

Squatter settlements in which the community mortgage programme is to be implemented

Panakan

Samal Island

Davao Gulf

Centre of City Poblacion

Davao River

Dumalag Point

Daliao

0   3 miles

0   3 km

government. Considering the size of these settlements, the high educational standards of many of the residents and the fashionable notion of 'people power', the Community Mortgage Programme appeared to many to be unnecessarily bureaucratic.

The experience of Davao City can be replicated throughout many cities in the developing world – in the *barriadas* of Peru, the *colonias proletarias* in Mexico, the *favelas* in Brazil, the *bidonvilles* in French-speaking countries, and the *bustees* in India. From a government's point of view, however, the problem is not just one of housing. There is also the problem of trying to provide clean water, proper sewerage systems, electricity, properly constructed roads and many other services, including health clinics.

Pollution

In writing about the economic success in East Asian cities, Wade has noted that 'anyone who has experienced the pollution and congestion of East Asian cities will realise that "success" has to be used in a qualified sense' (Wade 1990: 69). Certainly air pollution is an almost universal problem, though it is to be doubted whether the levels of pollution in Third World cities are any worse than those experienced today in, for instance, the cities of Poland and other parts of Central Europe. It is also arguable whether pollution is any better in the overcrowded, insanitary huts with cramped cooking facilities, charcoal fire and kerosene stoves characteristic of most rural settlements in developing countries.

As for the global emissions of CO_2 by region (Table 5.3), the latest figures show the following: Asia, 37 per cent of total global emissions; North America 27 per cent; Europe 19 per cent; Russia 8 per cent; South and Central America 5 per cent; Africa 3 per cent. In the 1998 UN conference on global warming in Buenos Aires, only developed countries agreed to binding targets for emission reductions. Developing

Table 5.3 *Carbon dioxide emissions (metric tons per capita), 1995*

	CO_2 emissions
High-income countries	12.5
Europe and Central Asia	7.9
Middle East and North Africa	3.9
Latin America and the Caribbean	2.6
East Asia and Pacific	2.5
South Asia	0.8
Sub-Saharan Africa	0.8

Source: World Bank (1998b)

countries insisted that today's problem is largely due to the excesses of the rich world: the United States alone is responsible for about a quarter of the world's carbon dioxide emissions. Governments from the developing world point out that cutting emissions in their countries would come at the price of economic growth, making the life of the poor even worse. On the other hand, it is the developing world that will be producing the lion's share of global emissions within a few decades. The Americans insist on 'meaningful participation', at least by the larger developing countries, but countries such as China and India have so far refused even to discuss the matter.

But perhaps the most serious environmental problem of pollution in the developing world concerns water, a problem made worse by inadequate sewage disposal systems. Many urban dwellers obtain water from standpipes, polluted rivers and water vendors. In Jakarta, for example,

> 40% of households are dependent on water vendors and have no waterborne sewerage system, and only 15% of the city's population have access to the city's water supply. Shanghai, which is now expanding very rapidly, is facing a real problem with its water supply and wastewater disposal (Ward and Liang 1995). In Manila, the sewerage system built in 1909 for 440,000 residents is now the only facility for 6 million people.
>
> (Potter and Lloyd-Evans 1998: 197)

Unemployment

In most poor countries, rates of urban unemployment are between 10 and 20 per cent of the total active population, but these data do not usually include registered members of the informal sector as unemployed, so that in practice the real figures are much higher. It is indeed difficult to identify the unemployed, and it is for this reason that some authorities believe that the term 'underemployment' provides a better indication of the true employment problem (Gilbert and Gugler 1992).

In discussing urban poverty in Africa, Potts (1995) draws attention to the impact of the IMF and its structural adjustment programmes (SAPs), which, she argues, 'have combined to devastate the real incomes of many urban dwellers in Africa'. She believes that the rate of urban growth in some African countries has slowed, and in some cases this has even led to 'reverse' migration.

Unemployment – the lack of well-paid employment opportunities for the majority of urban dwellers – is of course a central and serious problem, with its dangers of urban unrest, the loss of valuable human resources to the economy, and the social and environmental problems to which it gives rise. Potter and Lloyd-Evans (1998) have summarised the four major general characteristics of urban unemployment. These are as follows:

1 The urban labour forces in developing countries have grown at high rates of 2 per cent a year, representing both population growth and urban in-migration.
2 Rural–urban migration has overstretched the coping mechanisms of most cities, leading to an informal sector.
3 There is a close relationship between levels of unemployment, poverty and inequalities in the distribution of income.
4 Urban labour markets are shaped by their role in the wider national and international economy.

Urbanisation and development

The above brief discussion of some of the problems associated with urbanisation and its associated industrialisation has tended to produce a rather negative view of urbanisation and city growth in the developing world. Certainly many of the problems seem to be insurmountable, and they are often set against a possibly over-romanticised view of rural life in poor countries. Mention has already been made of the criticisms inherent in 'urban bias'. There is also a political angle to consider, for many neo-Marxists view cities as agents of imperialism and external neo-colonialism. Even when some degree of urbanisation is accepted as inevitable, it is argued that governments should do all they can to slow down the rates of urbanisation and avoid the problems associated with, for instance, over-urbanisation. There is a need for a proper balance between rural and urban incomes and for changes in government policies. Integrated rural development programmes should reduce unnecessary economic incentives for rural–urban migration. But should governments attempt to curtail the flow of migrants to the cities – usually to the largest cities?

In unstable countries there is often a conflict between town and country which makes the improvement of living standards in both almost impossible. 'While urbanisation does have obvious attractions, the social costs of a progressive overloading of housing and social services, not to mention increased crime, pollution and congestion, tend gradually to outweigh these historical urban advantages' (Todaro 1997: 265). Moreover, urbanisation is usually at the expense of small or medium-sized towns; for it is to the large towns and cities of developing countries that most rural–urban migrants go. For this reason large towns and towns tend to grow more quickly than small and medium-sized towns. Many writers also focus on the negative aspects of the role of cities in national integration and in world communications. In practice their role in, for instance, finance is very limited. For only three world cities really count in this regard: New York, London and Tokyo.

On the other hand, many authors consider urbanisation and economic development to be virtually identical processes; they believe that city growth is a necessary accompaniment to development. In Friedmann's core–periphery model, in which the urban city region centre organises the space economy, cities and city regions are regarded as the catalyst of growth. The advantages of city growth are well recognised. They 'offer the cost-reducing advantages of agglomeration economies and economies of scale and proximity as well as numerous economic and social externalities (e.g. skilled workers, cheap transport, social and cultural amenities)' (Todaro 1997: 264). The relevant social transformations can be more easily effected in towns than in rural areas. The negative features of urbanisation need to be viewed as problems to be solved, not as reasons for controlling or reversing the process. Urbanisation can in this sense be regarded as a panacea for the ills of developing countries. If the proportion of people living in cities increases as it is doing today, then clearly what happens in towns and cities is central to development as perceived by the people.

The negative view of urbanisation in relation to development in the developing world should not be rejected too easily. While it is true that there is strong evidence to support the view that urbanisation and development often proceed in tandem, this is not always the case, and the problems of urban pollution, congestion and poverty can in certain circumstances inhibit development.

As for the positive view of the role of urbanisation in development, from the economic point of view urban industry provides items for EOI as well as centres for the distribution of research, machinery and rural development. Rapid urban growth provides a burgeoning market for agricultural produce; and urbanisation can best accelerate the industrialising of agriculture. Socially, urban centres can best provide educational, health, literacy and communication services, as well as clean water and good public utilities. Urbanisation may also reduce population fertility. Finally, urbanisation widens the contacts and range of interests of the population, putting them in touch with the wider world. And politically, urbanisation makes political education easier, party politics more effective, opposition to the government more likely, and participation in the political process more complete.

Summary

The main focus in this chapter is on the problems, processes and policies of urban-industrial development and how these link into and affect the nature and extent of development. After a review of the levels and dynamics of urbanisation, including rural–urban migration, in developing countries, a brief section comments on how far there is a distinctive Third World city. The discussion then addresses and evaluates the informal sector, housing squatters and shanty towns, pollution, and unemployment. While there are clearly many negative features about urban-industrial development, there are, equally, many positive features. It is not surprising that a number of writers consider urbanisation and development to be virtually identical processes, though whether this is always true of urbanisation in developing countries is uncertain.

Questions for discussion

1 Distinguish between the formal and informal sectors in cities of the developing world. Identify the positive and negative features of each.
2 Identify the major causes and effects of rural–urban migration. Illustrate your answer with reference to one major city in the developing world.
3 How far do you agree that urbanisation, industrialisation and development are closely linked processes in Third World cities?

Further reading

Afshar, H. (ed.) (1996) *Women and Politics in the Third World* London: Routledge.

Cohen, A. (1971) *Custom and Politics in Urban Africa* Berkeley: University of California Press.

Drakakis-Smith, D. (1995) 'Third World cities: sustainable urban development – 1' *Urban Studies* 32: 659–77.

Friedmann, J. (1986) 'The world city hypothesis' *Development and Change* 17: 69–83.

Gilbert, A. (1996) *The Mega-city in Latin America* New York: UN Press.

Gugler, J. (1997) *Cities in the Developing World: Issues, Theory and Policy* Oxford: Oxford University Press.

Harris, N. (1992) *Cities in the 1990s: The Challenge for Developing Countries* London: UCL Press.

McGee, T.G. (1994) 'The future of urbanisation in developing countries: the case of Indonesia' *Third World Planning Review* 16: iii–xii.

O'Connor, A. (1991) *Poverty in Africa: A Geographical Approach* London: Belhaven.

Potter, R.B. and Lloyd-Evans, S. (1998) *The City in the Developing World* London: Longman.

White, R.R. (1994) 'Strategic decisions for sustainable urban development in the Third World' *Third World Planning Review* 16: 103–16.

6 ▸ The state context: China and Brazil

- The state and development
- Development planning and structural adjustment programmes
- The future of the state
- The case of China
- The case of Brazil

The state and development

Although every state is involved to some degree in development within its borders, there are still some countries in which development is viewed as best directed, controlled and planned by the central government. But by far the majority of states have now embraced the market as the engine of growth, one in which the role of government, in theory at least, is reduced. Even the present-day Russian Federation now practises market economics; and in China, while the government retains firm central political control, it is allowing an increasingly open, indeed capitalistic, economy to develop.

It would seem almost axiomatic that the geographical boundaries of a state should be the main context for a country's development planning. On the other hand, the external forces, including globalisation, operating on a country's economy, of whatever ideology, are always significant and, in an increasingly globalised world, may become dominant. It is also true that these external forces may be channelled through regional groupings of states, but these matters will be considered in Chapter 7. In the present chapter the emphasis is on internal development, either by government, the market or, in some cases, by internal regional activities.

The tendency to oppose government state control and the market economy too starkly is misleading and leads to false conclusions. It is sometimes implied that government intervention is characteristic only of command, communist countries, and that free-market economies are characterised by little or no intervention. According to some writers, governments in East Asia began to allow market forces to operate; progressive liberalisation and opening up of the economies to outside markets have subsequently ensured success. Success in Japan, for instance, has been due to private individuals and entrepreneurs responding to the opportunities provided in free markets. The only useful role governments have performed is to provide supportive environments in which entrepreneurs are enabled to perform their functions efficiently. Friedmann's view on this is quite categorical. In arguing for reducing the role of government he claims that 'Malaysia, Singapore, Korea, Taiwan, Hong Kong and Japan – relying extensively on private markets – are thriving. . . . By contrast India, Indonesia and Communist China, all relying heavily on central planning, have experienced economic stagnation' (quoted in Wade 1990: 22).

But this is only one point of view, reflecting a particular perspective on development at one particular time. In all countries of the Far East, for instance, governments have intervened, often quite heavily. It has been pointed out that

> the view of Singapore as the archetypal laissez-faire economy could not be further from the truth. However this is not to say that it is not a capitalist country, but rather that its success has not been built upon the operation of the free market. The government has been highly interventionist and has very carefully, and very successfully, for much of the time, stage-managed Singapore's development.
>
> (Rigg 1991: 196)

Indeed, the question must be asked – why, in spite of all the earlier 'success' achieved by the economies of the Far East, did the East Asian miracle suddenly disappear? For by 1997 a crisis had overtaken all these 'successful' economies of the West Pacific Rim. Many explanations, including 'the convulsive nature of global capital movements' and 'a too complete acceptance of the American-style political freedom and economic liberalisation', have been suggested (Poon and Perry 1999). According to Baer et al. (1999), this crisis placed into high relief many institutional characteristics of these Far East economies which negate many past attempts to characterise them as open and market driven.

Clearly, it is false to make a precise distinction between planned or command economies and those that profess to be market economies. Indeed, the consequences of dichotomising choices between planning

Plate 13 *Voting in Nairobi, Kenya*

and markets in this way are serious both for theoretical analyses and for practical policy-making. In almost all economies the strategies are now mixed. This is logical in the sense that the present trend in thinking is to emphasise the mixed nature of economic development strategies.

In his book *Governing the Market*, Wade considers the polarised views on the role of government in the economic development of East Asia. At one end of the ideological spectrum are the neo-Marxists and dependency theorists who emphasise the need for government and socialist state institutions to create the necessary preconditions for successful and socially equitable economic development. At the other end there are those who adopt a neo-classical analysis and emphasise the need for relatively free and open market forces to operate. Wade argues that neither of these polarised views is valid. What needs attention in the analysis is the *way* in which governments allocated decisions: 'The government did not simply control markets; it also offered periodically updated visions of the appropriate industrial and trade profile of the economy and gave a directional thrust to private sector choices in line with these visions' (Wade 1990: 4). In other words, East Asian industrialisation has depended critically neither on laissez-faire development policies nor on central government control, but on selective, guided but extensive government intervention.

There is an inevitable tension between the state, the larger multi-state region and the global context of development thinking and action. The state focus is as great as perhaps it has ever been. But far-reaching developments in the global economy have us re-examining basic questions about government: what its role should be, what it can and cannot do, and how best to do it (World Bank 1997a). The past fifty years have shown both the benefits and the limitations of state action, especially in promoting development. Will state governments be able to adapt to the demands of a globalising world economy?

Four recent developments have given these worries special impetus:

1 The collapse of command economies in the former Soviet Union and Central and Eastern Europe;
2 the fiscal crisis of the welfare state in most industrialised, developed countries;
3 the important role of the state in the East Asian 'miracle' states up to 1997; and
4 the collapse of states and the explosion in humanitarian emergencies in several parts of the world.

A common view today is that without an effective state government, sustainable development, both economic and social, is impossible. But this does not mean that the state must be solely responsible for providing development.

Percentage of 1989 revenue

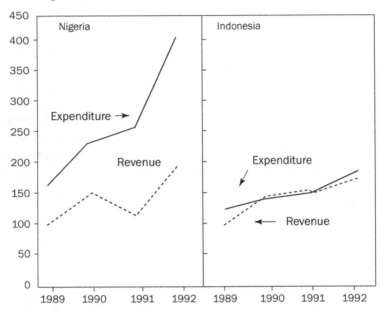

Figure 6.1 *Nigeria and Indonesia: contrasting reactions to oil windfalls*
Source: IMF, various years; World Bank (1997)

According to one's perspective on development, evidence now suggests that the state should be regarded as central to economic and social development, not as a direct provider of growth but as a partner, catalyst and facilitator. However, what is meant by an 'effective' state is not easy to determine. Even for countries at the same level of income, differences in size, ethnic make-up, culture and political systems make every state unique. Furthermore, in some countries governments have held back development through the use of arbitrary power, by corrupt behaviour and by destroying the confidence of private investors in the economy. In such countries – notably in parts of Africa – development has faltered and poverty has continued: how governments handle their resources can dramatically affect their development potential, as a comparison between how oil windfalls were handled in Nigeria and Indonesia respectively shows (Figure 6.1). Other problems have arisen with the failure of central or command economies, especially in the former Soviet Union and in Eastern Europe. This failure to deliver development was probably a main factor in the demise of communist governments in these countries. But in turn the collapse of central planning has created problems of its own, for in the resulting vacuum there is the danger that law and order deteriorate, basic goods are not available, poverty increases and instability becomes chronic.

Development planning

Development plans directed at change over periods of four, five or more years were at one time the *sine qua non* of development. Initially, at least, a development plan for a developing country was regarded as an almost required part of the process of attaining independence from the former colonial power. In practice, however, few of these development plans succeeded, and dependence on such international agencies as the World Bank and the International Monetary Fund (IMF) now dominates most serious development planning proposals. Inevitably, development plans today emphasise Western values and so devalue indigenous and traditional systems of knowledge, economy and culture. Development plans of this kind tend to discourage further independence or secession movements. The underlying philosophy of development planning today expresses a preference for the role of the market and capitalism against communism and state intervention.

This discussion raises the matter of structural adjustment programmes (SAPs). Beginning with a plan for Turkey in 1980, the World Bank and the IMF have been active in designing and supporting policy and institutional reforms aimed at a country's debt reductions and economic growth along the lines set out by the World Bank and IMF (Potter *et al.* 1999: 166–9). But SAPs have come in for a good deal of criticism. Dasgupta (1998: 315) has examined the SAPs imposed on the Third World and the former Soviet Union by the World Bank and the IMF, and finds far too little flexibility. He also points out that SAPs tend to ignore considerations of inequalities within a state, being essentially aspatial in concept.

It is suggested by some authorities that the World Bank, its affiliate the International Finance Corporation and the IMF do a great deal of damage to a developing country's economy in terms of encouraging dependence and destroying indigenous-based practices. But as access to international capital markets has dried up, firms and governments from the developing world have no option but to continue their dependence on the major international agencies for financial support. This is as true for the former Soviet Union as it is for Ghana.

Moreover, there are many examples of successful support from the World Bank and IMF. Egypt provides one good example in that the former statist, socialist Egypt has now become the very model of a modern emerging market, and the IMF's prize pupil. In Latin America Chile provides a most successful case. After the overthrow of the left-leaning president Salvador Allende in 1973 by a military coup, the main thrust was towards privatisation, mainly to undo the measures taken by the Allende regime to boost the public sector. The economy collapsed, and World Bank and IMF support has been the significant factor in the country's significant development since 1985. In Ghana after 1983, the military government religiously followed the reform programmes of the World Bank and IMF. But the aid has not reached all Ghanaians, especially those in the traditional agricultural industries (notably cocoa), which used to be the engine of Ghana's economy. Ghana is a relative success story, but again success has been selective in its impact. Another, less successful case is that of the Philippines, which, in spite of having the best growth potential in South-East Asia, has really been the region's sick man. The Philippines is close to the West in terms of culture and religion and, in terms of many social and economic indicators (schooling, education and the fact that over 50 per cent of the population is urban), is the most modern country in the region. But huge income disparities persist. The Philippines government has not been interventionist and has extolled the virtues of the free market. It is a regular client of the World Bank and IMF. But generous support from the World Bank and the IMF have failed to generate in the Philippines the kind of growth that has been witnessed elsewhere in South-East Asia.

The future of the state

The question being raised increasingly is whether the state is becoming an anachronism and whether the state as we know it has a future. Within the European Union this question forms one of the arguments presented against the development of the European Union as a future superstate in which the independence of its constituent governments might be progressively weakened. It is also true that the governments of states are becoming less interventionist and more global in their interests and outlooks. True isolationism is now impossible. Internally within states, too, as mentioned in Chapter 1, there is a tendency for the small-scale and local context of living and working to become increasingly significant. In a sense, therefore, the state context can appear irrelevant, being too small on a global scale and too large for local interests and networks. However, there is as yet no real sign of the decline of the nation-state as the natural context for development.

The case of China

In discussing the state context of development, the choice of China as a case study hardly needs any justification. It is easily the most populous state, is experiencing a satisfactory rate of economic development at the moment, and within a couple of decades is likely to be the largest economy in the world. It is also the only large remaining communist state, and as such is the outstanding example of a state whose economy reflects, in theory at least, strong central government control.

Some aspects of the Chinese experience in development have already been touched upon in earlier chapters, notably in discussions of population and migration, culture, and urban and peri-urban development. In a later chapter it will be necessary to discuss China in relation to world trading issues, notably the World Trade Organization (Chapter 7). In the present chapter, however, no attempt is made to provide a potted summary of China's economic development. Instead, attention is focused on five issues which need some further discussion and which relate to the varying experience of other states in different parts of the developing world. These five issues are:

1 the relationship between natural resources and development;
2 changing government policies in agricultural and industrial planning;
3 changing attitudes to commercial integration;
4 regional inequalities within the Chinese state; and
5 the Taiwan question.

Natural resources and development

It is commonly suggested that China is particularly well endowed with natural resources. A vast country with a wide range of contrasting environments, China also has a varied array of minerals and power resources for industry. However, it is easy to forget that the country's natural resources for agriculture – the mainstay for the majority of the population – are far from satisfactory. Indeed, China faces two fundamental ecological constraints which have deeply affected, and continue to affect, the country's agricultural development.

The first of these is the serious shortage of good cultivable or workable land. Although China is an immense and ecologically varied country, only 11 per cent of its territory is cultivated; moreover, it is widely accepted that the scope for increasing the cultivated area is negligible and that increases in production will have to come from the intensification of production on existing agricultural land. This is a particularly critical problem because of China's vast population of well over a billion. Intensity of land use is already very remarkable and the density of population per unit of cultivable land is exceedingly high. Over 90 per cent of China's population lives in the south-eastern third of the country. Land shortage is also partly a function of location, slope and soil. Much of western China is believed to be too inaccessible to be worth developing for agriculture, and in large areas of the country the land rises to great mountain ranges, high plateaux and hills, or descends into isolated interior basins. Steepness of slope is a severely limiting factor even in parts of the south-east, which is more mountainous than many accounts suggest, and where the terracing of land for agriculture is often necessary.

Soils, too, are generally poor or difficult to work, especially in the south, where they are frequently heavily leached. Levels of fertility in China are not naturally high and soils have traditionally demanded applications of nightsoil, animal manure, crop residue, ashes and mud, without which present levels of production could not be sustained. And even in parts of the north, where better soils – loess, chernozem and chestnut soils – are found, they often suffer from wind erosion, a lack of nitrogen and a serious lack of moisture.

The second major constraint facing Chinese agriculture is the broadly monsoonal climate, which results in a lack of water. Much of China north of the Yangtze River is a moisture-deficit area with less than 1000 mm of rainfall a year. Moreover, seasonality and variability of rainfall become more marked northwards and westwards, making even parts of the best land 'marginal' from time to time. Less than a quarter of China's cultivated area receives enough consistent precipitation for reliable agriculture, and great improvements in irrigation – largely tubewells in the north and canals in the south – and water control by dams and river diversion are essential requirements for intensive agricultural production. Temperatures, too, pose a serious problem in the north, where large areas have fewer than 200 frost-free days a year.

This issue of climate as a critical factor in agricultural development is recognised by the Chinese to a quite unusual degree in that most official as well as many academic sources make much of the effects of climate and weather on production in any one year. The apparent obsession in official sources with weather as a critical determinant of agricultural production in China is to some extent an admission that agriculture is still very largely weather dependent and that much remains to be done in water control, irrigation, scientific crop breeding and micro-climate control. But it is also a frank recognition of the serious climatic hazards – floods, droughts and typhoons – which have always plagued China's territory, more especially in the eastern river valleys, flood plains, river basins and deltas where so much of the country's best agricultural land lies.

The natural resources of China, then, can be broadly described as adequate for industrial development but difficult as far as agricultural development is concerned, particularly considering the huge population needing to be fed. But perhaps the main point to be made here is that China illustrates very well the lack of any simple causal connection between natural resources and development; it also emphasises the truth that in China, as in any developing country, the presence or lack of natural resources is never the reason for success or the excuse for failure in achieving development: 'the Creator has not divided the world into two sectors, developed and underdeveloped, the former being more richly blessed with natural resources than the latter' (Bauer and Yamey 1957: 46).

Changing government policies in agricultural and industrial development

The origins of many rural development policies in China lie well back in the period before 1949, when Communist control began; for it was in the Communist areas of pre-revolution China and during the war with Japan – especially in the Border Regions – that many of the characteristics of Maoist economics began; it was also during this period that the Chinese

Communists found it expedient to reject doctrinaire extremes and to adopt pragmatic and often moderate land redistribution policies, searching always for mass grassroots support. At this time, too, labour was used to replace capital, improved technologies supplied substitutes for imports, and local community initiative was substituted for centralised planning. As one authority has put it,

> in this way, by accident of history, the group of Chinese Communists most closely associated with the Border Region administration became expert in the Nurskian use of surplus labour, the use of intermediate technology, and the encouragement of community development, before these phrases were invented in the western world.
>
> (Perkins 1978: 561–6)

Moreover the settlement of refugees from the Japanese-held areas provided the Communist administration with the experience which enabled it to collectivise Chinese agriculture with so little disruption in the 1950s. But of all these institutions in which the Border Regions of this pre-revolutionary period anticipated later Maoist policies, by far the most important and striking were in origin non-communist: the Chinese industrial cooperatives. It was from these institutions that Mao took ideas for the Great Leap Forward, the beginning of the People's Commune, decentralisation, and the integration of non-agricultural services such as health and education into his concept of rural development.

Clearly, then, many of the initial policies, ideas, philosophies and institutions which are commonly associated with China after 1949 were in fact fashioned during the early struggles of the Communists before the revolution; and it is this fact that helps to explain many of the elements of China's rural development today.

During the period from 1949 to 1978 the government made many changes in agricultural and industrial policy. The first five-year plan (1953–7) saw the introduction of collectivisation and the adoption of the 'Stalinist' model of heavy industrial development. By 1957, however, it was apparent that the failure of agriculture to supply the agricultural raw materials and food for the non-agricultural, urban workforce was holding back China's industrial progress. Emphasis, therefore, was changed to focus on agriculture: the slogan 'agriculture is the foundation of the economy' was made central to Chinese economic policy.

This new initiative by Mao, encapsulated in the Great Leap Forward, included encouraging local initiatives in small-scale industry and local economic self-sufficiency. Furthermore, the division between industry and the large-scale urban economy on the one hand and rural life and the handicrafts of the countryside on the other was to be broken down. Central to change in this period was the growth of the People's Commune, which persisted even after the failure of the Great Leap Forward, the Great Depression (1959–61), Readjustment (1961–7), the Cultural Revolution (1967–70) and the Revival of Growth with Trade (1970–7).

After 1978, although many of the ideas, concepts and institutions existing before that date were to be incorporated into a series of reforms, there was nevertheless a marked and major change in the attitude of the Chinese government towards agricultural, rural and urban-industrial development which was to give the reforms a very different character from any policy changes that had occurred previously. There was, in particular, a deliberate movement away from those forces tending towards a fragmented, cellular economy. The emphasis now shifted towards the formation of an integrated national commodity economy. In practical terms Maoist communist ideology was adjusted to deal with a China that was hoping to develop economically and 'open up' much more than hitherto to the outside world.

Whatever success China's economy has achieved since 1978 must be put down in large part to changes in government policy. The government certainly remains communist, authoritarian and non-democratic. It is still in almost complete control of the economy, operating within a tightly controlled political and administrative system, and it seems determined to maintain China's position as the last great communist state. Yet in 1978 the government made a significant break with the past and introduced what it terms 'market socialism', the purpose of which is to modernise China's economy and improve the standard of living of its peoples. It has attempted to reinstate markets as a central feature of socialist economic theory and practice. The ownership system has been diversified, with private enterprise gaining some ground, especially in the commercial and service sectors. But state ownership remains dominant, even though decisions and control of operations may lie with the individual.

More generally, the government has been moving from a policy of import substitution industrialisation (ISI) to one of export-oriented industrialisation (EOI), though cautiously and slowly. Furthermore, there is renewed emphasis on agriculture in an attempt to deal with the need for improved food supplies, especially to the urban-industrial areas, and to increase the purchasing power of the rural population. The government has also encouraged a shift away from heavy to light industry. But perhaps the most important change has been the government's 'open-door' policy of encouraging foreign direct investment (FDI) and technology, together with the setting up of Special Economic Zones (SEZs) into which foreign investment and semi-capitalist or capitalist enterprises have been channelled. During its reform period China has experienced dramatic changes in industrial ownership structure. The proportion of industrial output produced by state-owned enterprises (SOEs) has declined, while that of non-state enterprises has increased rapidly (Wei 1999). Coastal regions have experienced especially rapid transformation of ownership compared with the western provinces. But SOEs still dominate in the traditional industrial bases in many interior provinces.

The changes in government policy towards agricultural and industrial development since the 1949 revolution very briefly discussed here have been numerous and often conflicting. Some have been disastrous (like the Great Leap Forward and the Cultural Revolution), but others have helped to bring about a less fissiparous and more integrated economy and society on which the present reforms can now build.

Commercial integration

A condition of commercial integration applies not just to physical inputs and outputs for the production and marketing of consumer goods: it extends throughout a society. Indeed, commercial integration is often taken to be synonymous with economic development. In Western Europe, certainly, national commercial integration and machine technology are generally regarded as having been the core of the Industrial Revolution, affecting all sectors of life: private and public, domestic and foreign, the family and the individual, the Church and politics.

As perceived in China, commercial integration is undoubtedly the goal towards which the reforms introduced after 1978 are being directed. The Chinese are well aware that doubts have been expressed about the practicability of encouraging commercial integration while at the same time adhering to communist ideology and enabling the state to retain firm control over the polity and economy of China. However, some writers argue specifically that one reason for the failure of China to achieve more rapid economic growth or to have progressed further towards the ideal communist state is that a capitalist stigma has wrongly been attached to the notion of commercial integration. Provided that the current structure of ownership remains intact and profits are used to further the economy and welfare of the state and all working people, commercial integration is now viewed in China as being entirely consistent with the economic development of a communist state. The theoretical arguments explaining this consistency between commercial integration and communist ideals are found widely in Chinese-language sources (but see, especially, a summary in English by Zhou Shulian and Wang Haibo, 1982).

Commercial integration, so it is now believed in China, can and must be achieved without its historical corollary: capitalism. But whether this is little more than thinly disguised rationalisation remains to be seen.

Regional inequalities

The regional differences within a state in levels of development can be regarded as undesirable if they threaten the unity of the state and encourage centrigugal forces to become too strong. Regional inequalities are no less striking in China than in many other countries (Figure 6.2). In particular, and resulting largely from China's open-door policy, the two southern provinces of Guangdong and Fujian are now experiencing the kind of economic growth that has attracted the description 'miracle'. The pattern of industrial development and the regional differentials in growth rates in productivity and prosperity seem likely to create stresses in the unity of the state. The Beijing government is of course well aware of this problem and is in various ways trying to redress the balance between the booming south and the rest of the country. An attempt is being made to reinstate Shanghai as the country's major financial and business centre. Shanghai is receiving support for its development projects, notably the Pudong Development Project just to the east of the city. This involved developing 350 square kilometres of farmland and swamp into an industrial, commercial and financial zone. But perhaps the greater dangers relate to the great contrasts in prosperity between the coastal and inland regions of China. China's government is now struggling to mitigate the disparities that have arisen as a result of the reforms. As Figure 6.2 shows, the coast has far outpaced China's spawling interior. The gap is vast, in terms both of personal income and of economic output. Among the poorest areas are those populated by the non-Han minorities such as Tibetans, Muslims (the Hui) and Mongolians. Attempts to redress regional imbalances include incentives to operate in the interior, to shift factories inland and to invest inland. But there are many cases of local protectionism. Governments at provincial, county and city levels routinely block goods from other regions. Sometimes tariffs are imposed, but more often the barriers are bureaucratic.

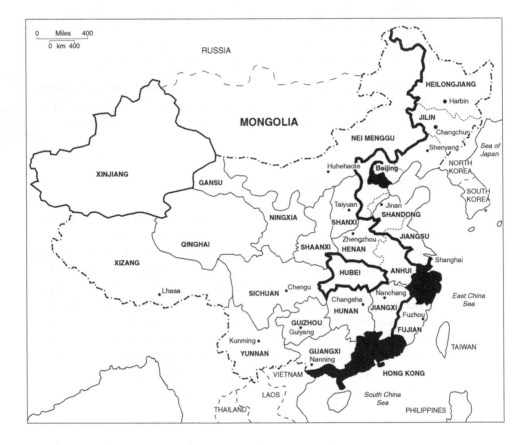

Figure 6.2 *Regional income inequalities in China, 1997*
Source: Adapted from *The Economist*, 16 January 1999, p. 68
Note: The three areas in black have GDP per capita of over 10 000 yuan. Xinjiang, Hubei and all the coastal provinces inland to as far as the thick black line have GDP per capita of 5 000–10 000 yuan. The rest of China has GDP per capita of less than 5 000 yuan

China is a good example of a country where any analysis of its economic development, either current or future, must be underpinned by political considerations. This not just because it is a communist country, with communism's predictable theory and practice, involving politics at every level of its economic and social life. It is also because China is now the last great communist state, and it cannot avoid looking over its shoulder at what has happened to the former Soviet Union and the countries of Eastern Europe. The Chinese government seems to be determined that China will remain a communist country and that its state identity will remain intact. There is to be some *perestroika* but no *glasnost* in China. While the government is ready to consider limited and controlled liberalisation of the economy, such liberalisation must always operate within the limits set by the political imperatives.

The Taiwan problem

One important issue which arises in any discussion on China and must be given brief mention is the question of Taiwan, or the Republic of China as it is formally known. China claims Taiwan as a legitimate part of its sovereign territory and will not recognise the island as a separate state. Taiwan had its origins as a separate 'state' in the Communist takeover in China and the consequent flight of the Nationalist government and its supporters to what was then known as the island of Formosa. Underlying the economic success story of Taiwan is its government's determination to show that its economic policies are more successful than those of the centrally planned and controlled communist state of China. While Taiwan's economy is based on a free-market, capitalist system, its planning is designed and carried out by government. Certainly this success has been dramatic, resulting in annual per capita GDP growth rates of 7 per cent over many years; between 1984 and 1989 the per capita income doubled. Taiwan also has one of the lowest income disparities in the world.

The country's population of over 21 million, boosted by the huge inflow of immigrants from the mainland at the end of the 1940s and in the early 1950s, speak several different Chinese dialects, but Mandarin is now widely spoken in this highly educated country. Another important unifying factor among the population is their opposition to the Communist government across the straits in their homeland.

Looking to the future, it seems likely that Taiwan will attempt to bring about closer political and economic ties with Beijing. China is the natural source of labour and is the natural market for Taiwan's products, given the common culture and language. Improvements in Taiwan–Chinese relations is important to Taiwan, not only for economic reasons, but also because Taipei wishes to reduce international tension by removing any military threat from the mainland, and wishes to be accepted into the international community. As a diplomatic pariah Taiwan is at present banned from international agencies such as the IMF as well as suffering from the sense of isolation its present status encourages.

* * *

Some writers suggest that China's main problem with achieving development lies with the present political system – that unless China embraces democracy it cannot continue to prosper economically. Much of the literature in the West suggests that economic liberalisation and a repressive, totalitarian political system are incompatible. In China, however, there are many younger academics who see no logical reason why this should be so; and present evidence seems to support them. The problem is rather whether the Chinese government is prepared to adapt the nature, degree and style of its intervention so that it guides rather than rigidly controls the economy. There is plenty of evidence to suggest that economic imperatives as well as current theoretical and ideological arguments in China are being used to justify this kind of change or adaptation.

The Chinese no longer argue that communism and capitalism are logically incompatible. It is recognised that planned and market economies each have their own strengths and weaknesses and that China needs to draw upon the strengths of both in order to build a mixed, commercially integrated economy. While political imperatives may reduce the speed at which development occurs, there is no reason to believe that the dead hand of the state will continue to be anything like as damaging as some writers suggest. Even if one assumes the continuance of the present government, it is clear that this need not be an insurmountable barrier to growth. State industry is being reformed from within, but in a way and at a pace determined by political imperatives. Domestic private industrial enterprises are boosting the country's economy, helped incalculably by the inflow of investment, firms and technology from abroad. The economic modernisation of China has included investment in the country from outside. Over the past decade $200 billion of foreign investment has come to China. Foreign ventures now account for over half of China's exports, and the private sector now makes up one-third of the economy. Moreover, some state companies are now acting like private companies, and the government is facing up to the difficult social and economic implications – notably unemployment – of laying off much of the state sector. Development is also evidenced by the fact that 1.5 million new fixed telephone lines are laid each month, and China is already the second largest mobile phone market in the world. Private housing is catching on. And the one-child policy and the work-unit system are being adapted to current realities.

It may indeed be that the worst-case scenario for China's economic future is not that communist control will continue, but that the clamour for political freedoms will ignore and then destroy – as they have in the former Soviet Union – the economic progress on which the prosperity as well as the stability of the Chinese state must ultimately depend.

The case of Brazil

For all its superficial similarities with the rest of Latin America and the shared experience of European colonialism, Brazil is in many ways a quite distinctive country, held together yet isolated from the rest of the continent by its Portuguese language, its distinctive culture and its continental scale (*The Economist*, 27 March 1999). Brazil, which covers much of the Amazon Basin and the great Brazilian plateau, covers almost half of the area of South America; it is also the world's fifth-largest country. It is also distinctive in being the only truly federal country in the region, with the Federal District of Brasilia and 26 states. Dominant in Latin America, Brazil has an abundance of most natural resources and its prospects for successful development are in theory good, although it is sometimes cited as a country which suffers from the 'curse' of good natural resources, thereby lacking any stimulus to utilise fully its non-natural resources. On the other hand, Brazil does have problems with its natural resources, including the suggestion that it has too large an area for its population, that water supply, notably in the 'dry' north-east, is often inadequate or badly distributed, that in much of the country the climate is enervating, and that many of the most valuable mineral deposits, including iron ore, manganese, tin, bauxite, copper and gold, are so located that their exploitation is very expensive. On balance, however, it is difficult not to accept that the natural resource base of Brazil is of outstanding size and quality; it certainly cannot be used to explain human poverty or lack of development.

Land reform is regarded as an important issue in the country, for 50 per cent of the farmland is at present occupied by under 1 per cent of the farms. Using the Gini coefficient, indeed, Brazil's figure of 0.86 shows it to be one of the worst cases of inequality in landholding in the world. It is true that many of the largest holdings are in the forested areas of Amazonia and so are not suitable for agriculture; moreover, soils over much of Brazil are very poor indeed – some estimates put the potentially arable land at no more than 17 per cent of the total land area of the country. Some have suggested in the past that in Brazil the large holdings were necessary to satisfy the needs of commercial agricultural production and therefore that Brazil 'must unfortunately face a hard choice between equity and productivity' (Nicholls 1971: 387–8). Nevertheless, much has been done to bring about greater equity in the land tenure system, and it is now claimed officially that the whole institution of the latifundio is 'now defeated' in Brazil. In southern Brazil there are now many family farms, and the centre-west has a developing and efficient agribusiness, based on capital and technology inputs. This has developed in the *cerrados* savannas lying between the Amazon forest and the coastal states, where farmers from the south have migrated into the area to create a flourishing agribusiness of what is believed to be almost limitless potential.

Today Brazil is noted for some of the most innovative and successful attempts at land reform. One such example is in north-east Brazil (Piauí), where the World Bank has supported the Piauí Rural Development Project, involving land tenure reform. It shows that land transfers to small farmers are feasible and that security of tenure encourages poor farmers to adopt new technologies. Over five years the project bought 200 000 hectares, distributed land to 3 450 families and regularised land tenure for 1 500 families. The Bank also provided roads, water supply and irrigation. The total cultivated area increased by 16 per cent and the value of production by 50 per cent (World Bank 1998a: 65).

This is a significant development because in general, agricultural policies in Brazil tend to favour large farmers. The government's main successes in dealing with poverty are in the least favoured agricultural areas such as the poor, 'dry' north-east. Outside these areas agricultural policies in Brazil have reduced labour demands and have made it almost impossible for a poor person to buy land and become a farmer. Subsidised mechanisation has reduced labour demands and has reduced the possibility of unskilled labour acquiring mechanical skills.

With a population of 175 million, Brazil contains one-third of Latin America's population. The population growth rate is 1.9 per cent, the result of a rapid demographic transition, even though this is a largely Catholic country with little support for birth control: in 1988 each adult woman had 3.5 children, whereas today the figure is down to only 2. Over 65 per cent of the population lives in urban areas, reflecting to some extent the importance of industry in the economy. These urban centres have attracted large areas of slums and shanty towns, inluding the notorious *favelas* of Rio de Janeiro referred to in Chapter 5. Urban unemployment is a serious problem and the informal sector is large: in São Paulo it comprises 43 per cent of the total population.

Much outside attention has been directed at the plight of the indigenous people of Brazil. These account for only 0.2 per cent of the total population and their lands for about 12 per cent of Brazil's total land area. Land confiscation and natural resource exploitation, including forestry, have led to their position becoming increasingly precarious. Those invading the indigenous people's land are mostly marginal workers engaged in illicit activities such as gold mining or cutting down luxury woods. Other invaders are public-sector workers building new roads or hydroelectric stations. Particular concern is expressed about hydraulic gold mining when this is practised not by highly capitalised operations but by thousands of small-scale producers; this has resulted in serious sedimentation and health problems in downstream communities (Johnston *et al.* 1998: 327).

The two fields of policy which most observers believe most need tackling in Brazil are education and health. But education levels are improving and are now probably at levels similar to those in Argentina and Chile: 96 per cent of children under 14 are now in primary schools; and the illiteracy rate for the over-15-year-olds is now 19 per cent. As for health, the infant mortality rate is 58 per thousand and about 7 per cent of children are malnourished. A major contributing factor here, however, is the lack of sewerage systems, electricity or clean water, especially for the poor in urban areas. Beyond the urban core, conventional sewerage systems, with average household costs ranging from $300 to $1 000, are too expensive for most developing countries. But Brazil has been at the forefront in developing technological alternatives, including the innovative condominial system developed in the urban centres of the north-east. For a variety of reasons – high housing densities, impermeable soils and the need to dispose of a great deal of domestic waste water – on-site solutions do not fare well in many urban areas (World Bank 1992: 107–8).

As for poverty, Brazil has a high HDI rating, but the averages on which such statements are based cover enormous extremes. Brazil produces 43 per cent of Latin America's GDP, but almost one-third of Brazilians (50 million) are officially described as 'poor'. It is estimated, indeed, that 63 per cent of the total income in Brazil goes to the richest 20 per cent, while only 2.5 per cent of total income goes to the poorest 20 per cent. In geographical terms, the modern coastal cities and many areas of the south-central part of the country are reasonably prosperous, whereas the moisture-deficit north-east states are very poor, exhibiting all the characteristics of a poor Third World country. Brazil, indeed, has the most unequal income distribution in Latin America, which itself is the most unequal sector of the developing world.

Why Brazil should have such gross inequalities in income is an interesting question. One of the reasons must lie in its history and the unequal landholding pattern it inherited from its colonial period, together with the four-centuries-long history of slavery, the consequent neglect of education and the persistent culture of violence. More recently one must cite industrial protectionism, the recent history of chronic inflation, together with the debt crisis of 1982–92, which led to what is now called the 'lost decade'.

Brazil's income inequalities are very clearly expressed regionally. Income per head in the Federal District is seven times higher than in Piauí or Maranhão, the poorest states in the north-east. Since the 1960s Brazilian governments have been trying to reduce that gap, using regional policy and revenue transfers between states. In practice, however, this has meant trying to spread wealth and industry away from São Paulo. With 34 million people and an economy worth $280 billion in 1998, São Paulo State is not only Brazil's industrial heartland but also its main financial centre and an agricultural powerhouse. The gap between the nine north-eastern states and the rest of Brazil may be narrowing, but it remains large. Worse, the gap within the north-east between its three large cities – Fortaleza, Recife and Salvador – and most of the interior is also widening. Meanwhile, metropolitan São Paulo is going through a painful economic transition as its industries suffer from the effects of technological change – unemployment and severe social problems, including violence. The government's main tools of regional policy today are tax breaks – investments in the north-east and in the Amazon have been largely exempt since the 1960s – and revenue-sharing between states and the federal government.

Economically, Brazil is Latin America's giant, with the eighth biggest economy in the world, and is clearly dominant in the subcontinent. The case of Brazil was originally widely used by early theorists, notably Frank, to illustrate the 'development of underdevelopment' thesis referred to in Chapter 1 (Frank 1967: 30). The history of Brazil is perhaps the clearest case of both national and regional development of underdevelopment. The expansion of the world economy since the beginning of the sixteenth century successively converted the north-east, the Minas Gerais interior, the north and the centre-south (Rio de Janeiro, São Paulo and Paraná) into export economies and incorporated them into the structure and development of the world capitalist system. Each of these regions experienced what may have appeared as economic development during the period of its golden age. But it was a 'satellite development which was neither self-generating nor self-perpetuating' (Frank 1967: 31). Brazil's more recent economic history, however, reflects changes in economic and social strategies in Latin America. Brasilia was an attempt to join together the constituent ('satellite') parts of the country and so to encourage a sense of unity. President Fernando Enrique Cardoso, known originally as an influential theorist whose early work was anchored firmly in the dependency approach, was elected on a platform of economic and trade liberalisation. The country's policies are now characterised by the widespread use of 'targeted programmes' – targeted, that is, on specific groups of people. In other words, policy is moving away from state controls and public investment towards open economic systems relying on private investment and market forces (Sheahan 1998: 93). Nevertheless, the state continues to play a dominant role in the country's economy, with the state-owned sector still accounting for almost 45 per cent of the country's GNP.

Brazil is one of the newly industrialising countries (NICs), and certainly industry remains at the heart of the country's economy, although geographically it has hitherto been concentrated in the south-eastern states of Rio de Janeiro, São Paulo, Paraná and Rio Grande do Sul. Now, however, it is growing in the north-east and far west. Today industry accounts for 35 per cent of GNP and 60 per cent of exports. The industrial base is broad, including steel, chemicals, petrochemicals and consumer goods. Brazil has an impressive hydroelectricity industry and is rapidly developing its nuclear power.

Agriculture

Agriculture in Brazil occupies 26 per cent of the population, accounts for 11 per cent of GDP and provides 40 per cent of the value of exports. Brazil is still the largest coffee exporter and the second largest cocoa and soya bean exporter, but produces a wide range of other export crops, including sugar and cotton, as well as meat. As the country has few petroleum reserves, it has gone a long way in developing an oil substitute from ethyl alcohol, extracted from sugar-cane.

One major problem in Brazil has been how to get farmers to accept new technology. One outstanding successful example is on the Formosa irrigation district in Brazil's north-eastern state of Bahia. When the project started, farmers in the local water-user association were reluctant to adopt efficient water management options, such as water-saving sprinkler systems and high-value crops. Water charges did not cover operation and maintenance costs. The system as it stood was unsustainable. In 1955, however, the reasons for lack of interest were identified and led to a change in emphasis, involving the farmers' children. In what was called Project Tomorrow, a vocational school was founded to teach the younger generation about better irrigation, new agricultural techniques and plant nursery management. The school also offered classes in sewing and other household skills. The young people have begun to persuade their parents to try new technologies and to plant high-value crops.

Viewed as a whole, Brazil's economy has recently (1999) been in some turmoil, leading to devaluation, recession and a great deal of pessimism among observers. But the agreement with the IMF reached in March 1999 now gives some grounds for hope, although the austerity policies recommended and imposed by the IMF have led to great hardships, with some of the regional states defying Brazilian state policy. If Brazil goes under, the repercussions would be continent-wide. The IMF has therefore applied conventional conditionalities to the loan, including a tight fiscal squeeze. Devaluation was caused ultimately by investors losing confidence, for whatever reason, in the government's ability to tackle the country's main problem, fiscal profligacy, and Brazil provides yet another example of the limited extent to which outside credit can shore up investor confidence. Nevertheless, the IMF has now agreed a $41.5 billion package to help Brazil maintain its drive for development, and this should be helped by the rapid demographic transition in Brazil referred to earlier. But there can be no successful improvement in the country's economy until the government can convince investors that Brazil is a safe place for investment.

Clearly, Brazil's economy faces many problems – of inflation, debt, foreign investment and declining exports. In 1994 the government introduced the *Real* Plan, which was an attempt, through prudence and austerity, to improve the prospects for development in Brazil. And the current round of World Bank/IMF funding should enable the Brazilian economy to recover before too long. But in the outside world the problem of environmental damage has attracted perhaps the greatest attention – notably that related to the destruction of the tropical rain forests, the largest rain forest in the world and covering about half of the country. Brazil is well aware of this issue and hosted the UNCED conference in 1992. There is now international cooperation to protect the Amazon. It is accepted that the Brazilian Amazon area is a unique repository of natural reserves of value to the world at large. The plan has brought together several federal agencies, the nine state governments of the Amazon Region, and numerous local and national NGOs. The Brazilian Tropical Rainforest Fund illustrates what can be achieved when donor and recipient countries cooperate to tackle the most urgent problems of preserving biodiversity.

Summary

This chapter considers first the state context of development – a context which is being increasingly challenged both from within states (for instance, by devolution movements) and from outside by regional groupings of states and by the processes of globalisation. The role of government in a state's development is crucial but it varies significantly in nature and in degree from case to case. On balance, there is a presumption that the state is likely to remain the most effective geographical area within which to plan development. The case of China is used to illustrate many of these points, but particular attention is paid to four issues in China's development: natural resources, changing government policies, attitudes to commercial integration, and problems of regional inequalities and the Taiwan question. In the case of Brazil, emphasis is laid on land tenure and agriculture; education and health; income inequalities; financial problems such as debt, investment and declining exports; and environmental damage.

Questions for discussion

1 To what extent do you think that the role of government is declining in most developing countries?
2 Why is it not always easy to establish market economies in developing countries? Give examples to support your viewpoint.
3 Identify the main obstacles to more rapid development in (1) China and (2) Brazil. How best can these obstacles be overcome?

Further reading

Becker, B.K. and Egler, C.A.G. (1992) *Brazil: A New Regional Power in the World Economy* Cambridge: Cambridge University Press.

Binns, T. (1993) *Tropical Africa* London: Routledge.

Drakakis-Smith, D. (1991) *Pacific Asia* London: Routledge.

Findlay, A.M. (1994) *The Arab World* London: Routledge.

Gilbert, A. (1990) *Latin America* London: Routledge.

Hodder, R.N.W. (1993) *The Creation of Wealth in China* London: Belhaven.

Leeming, F. (1992) *The Changing Geography of China* Oxford: Blackwell.

Perkins, D.H. (1989) *Asia's Next Economic Giant?* Seattle: University of Washington Press.

Wade, R. (1990) *Governing the Market: Economic Theory and the Role of Government in East Asian Industrialization* Princeton, NJ: Princeton University Press.

World Bank (1997) *World Development Report 1997: The State in a Changing World* Oxford: Oxford University Press.

7 ▸ Regional groupings, trade and aid

- Regional groupings of states
- Trade and development
- World trading regions
- Protectionism
- The World Trade Organization
- China and the World Trade Organization
- Aid

Regional groupings of states

The logic behind attempts to deal with state problems of development by grouping national economies together in a regional association seems unexceptional. As the case of the European Union (EU) has demonstrated, this approach applies equally to developed and to developing countries. In some, but not all, cases these regional groupings for economic cooperation are expressed in terms of trade blocs, to be considered later in this chapter. In some cases, too, regional groupings have arisen out of a perceived need for political cooperation or union, to confirm post-colonial independence from a former metropolitan power, for river basin development or to enable groups of states to have more weight in international affairs than if they acted alone.

The arguments supporting the setting up of regional groupings of states for the purposes of economic cooperation reflect many of the problems faced by numerous states in the developing world: small populations, awkward shapes, landlocked locations and an apparent lack of social, political or economic viability.

There have now been many attempts at regional cooperation, and perhaps the most successful to date has been the EU. But elsewhere – and even in Europe to some extent – it is fashionable to argue that successful regional cooperation between states for regional economic political and social cooperation will always founder on the rock of nationalism. However, the problem is much more complex than this, as can be illustrated by the case of West Africa (Box 7.1)

Box 7.1

Regional grouping in West Africa

West Africa is a good case study of the issue of regional cooperation because there would appear there to be a strong, compelling prima-facie case for some sort of economic, if not political, cooperation between the fifteen states of West Africa in what is known as ECOWAS (the Economic Community of West African States), established in 1976. It is a logical area for such development because in no other part of the world, except perhaps Central America, is there such a large number of states, varying greatly in size, but most of which have broadly similar economies and apparently little economic basis for independent political existence. Furthermore, three of these states – Niger, Mali and Burkina Faso – are landlocked, and so have to trade through and use port facilities in one or more of their neighbours.

Another reason for encouraging some form of cooperation between states in West Africa is to facilitate interstate trade beyond that demonstrated by the needs of the landlocked states. Interstate trade within the region could be an important means of widening the domestic market because, with the possible exception of Nigeria, all West African states suffer from the fact that their domestic markets for their own products are too narrow and limited.

It is true that this question of widening the domestic market in West Africa is not solely a matter of reducing the economic barriers set up by the present patchwork of political units; it is also related to the improvement and extension of transport facilities and to an increase in the purchasing power of the local population. Nevertheless, there is little doubt that the present pattern of international boundaries still works against the expansion of local markets, encourages the building of high tariff walls and, in general, helps to perpetuate the divisive tendencies in West African economic life.

Other factors suggesting the desirability for cooperation between West African states can receive only the briefest mention here: the need to reduce tariff barriers, and the opportunities for cooperation presented by the existence of ethnic communities spread across two or more adjacent political units. In economic development, cooperation is required to facilitate large-scale planning, to avoid unnecessary duplication in research, and to solve the practical problems of developing the major West African rivers, particularly the Niger, for hydroelectric power, irrigation and navigation. Politically, too, cooperation is necessary to increase West Africa's potential for effective defence – well illustrated recently in the use of the region's army in Sierra

Leone – and to increase the economic and political bargaining power of the region in continental and world affairs.

There are, however, severe difficulties in the way of successful regional cooperation in West Africa. Reference is often made to the fundamental dichotomy between north and south, in terms of environment, population, religion and economy; the complicated pattern of ethnic groups and languages; the different colonial associations, reflected in different official languages, educational and legal systems, and in administrative systems; the numerous currencies and trading relations; strongly contrasting levels of wealth and development; and the poor communications between neighbouring states. All these factors combine to hinder cooperation at any level. Then there are the extreme inequalities in size and population among the states of West Africa, which result in the small nations fearing domination by the larger. Moreover, there appears to be no deep-seated desire for regional cooperation. The concept of West Africa as a distinctive region is at best vague, and the exact form of any federation, confederation or customs union is no more agreed than is the method by which it might be achieved.

However, a number of attempts to form regional groupings within at least parts of West Africa have taken place since the late 1950s. One was the Benin–Sahel Entente. This entente between French-speaking states (Côte d'Ivoire, Burkina Faso, Niger, Benin and, later, Togo) was aimed primarily at economic cooperation between the member states. This attempt failed largely because of the great differences in wealth and levels of development between the member states. Côte d'Ivoire, while contributing so much more than other members, received relatively little and resented what amounted to subsidising its poorer neighbours. Other attempts have been made, but by far the most important and promising attempt at regional cooperation in West Africa is ECOWAS.

Source: Hodder (1978)

The problems of achieving any real or permanent measure of success in regional cooperation are very great and emerge clearly from the evidence of every case study. What is to be the practical basis upon which to build cooperation between neighbouring states? How can the richer states be protected from the feeling that they are being exploited in order to subsidise their poorer neighbours? Where are the actual development projects to be built? And what happens when member states feel that national interests are being overridden by those of the regional grouping? These are the questions that always have to be answered.

Of particular interest today is how to face the dangers of expanding existing regional groupings. For instance, the EU is aware of the pressures placed upon it by countries from Central and Eastern Europe that wish to join. There have been previous EU enlargements, raising membership from six to fifteen countries between 1957, when the six original countries signed the Treaty of Rome, and 1995; this involved countries more or less ready to join the EU at the time they applied, not those – including countries now classified as 'developing' – emerging from communist central planning and requiring fundamental reform, under EU supervision, prior to joining. Enlargement to the east will require a basic overhaul of the institutions and finances of the EU itself. The present system can cope with up to twenty countries. But enlargement beyond that number could trigger an intergovernmental conference to rewrite the EU's constitutional rules. Because of the problems of introducing a single currency, and the horse-trading of many issues among national governments that forms part of any EU negotiation, it is easy to see why enlargement has dragged on. However, the candidate countries are clearly prepared to be patient, for they have nowhere to go. They have the EU on one side. On the other is an 'arc of instability' sweeping down Europe's flank from Russia, through Ukraine, the Balkans and Turkey to the south and east Mediterranean.

Another kind of problem arising from expansion of a regional grouping is well illustrated by the current problems facing the Southern African Development Coordination Conference (SADCC). Set up in 1980 to reduce the region's economic dependence on apartheid South Africa, it later embraced majority-rule South Africa and in 1995 set about acquiring a new security and political role. But calls from Congo, which became a member in 1997, have brought about a split that threatens not just SADCC's security ambitions, but also its economic activities. The main problem is that South Africa is, in economic terms, nearly three times as large as the other thirteen members combined. It has a far more developed economy and most of its trade is with countries outside the region. Many members of SADCC fear that South Africa's huge companies just want them to produce commodities for South African industry, threatening whatever manufacturing capacity they may have built up themselves. As Zambia's then prime minister said when SADCC was set up, South Africa would always be an elephant surrounded by chickens. But in the case of Congo, South Africa was against sending troops in to help the Congo president, Laurent Kabila. However, it was forced to agree to it. Personal bitterness is now spreading and cooperation is diminishing. By admitting huge, unstable Congo, SADCC may have destroyed itself.

Trade and development

Some authorities argue that trade is the *sine qua non* of development, whether for the developing world or for the richer, developed world. In East Asia, for instance, it has been claimed that 'trade, trade and more trade was what propelled the Pacific Rim states out of agrarian destitution or post-World War II destruction and decline into world economic prominence' (Aikman 1986: 100). During the late 1970s and early 1980s the region's countries increased their share of European and North American industrial markets; and much of this trade was dominated by Japan, joined later by Hong Kong, Korea, Taiwan and Singapore, then by several of the ASEAN (Association of South East Asian Nations) countries – notably Malaysia, Indonesia and the Philippines – and finally by China.

The conventional wisdom about the trading problems of developing countries is that their economic development has been restricted by existing patterns of foreign trade. Trade between developing countries generally is said to be very limited: the range of their materials available for export is generally so small and so similar that they have very little to offer each other. International trade has been controlled and governed by the metropolitan, former colonial, powers and has been directed largely at those crops and raw materials needed by the manufacturing nations of those powers and by the urban industrialised world. At independence, developing countries were also left with a legacy of external trading organisations which restricted rapid economic advance. In particular, the association between colonialism and export economies was emphasised by the concentration of the large-scale export trade in the hands of a few large, mostly European, firms and by multinational corporations.

A further problem faced by developing countries at independence was that their exports to the industrialised world were largely of raw materials, in which they are supposed to have a comparative advantage, while their imports were manufactured goods from the developed industrialised countries. Primary, mainly agricultural, products and some minerals dominated exports, and finished consumer goods accounted for most imports. Furthermore, in most developing countries exports comprised only a very few items. Such a pattern of exports has repercussions upon the production side of the economy, on labour market policy, on soil fertility and conservation, on crop pests and diseases, and on the improvement of dietary standards.

However, there has been a marked change in the composition of exports from developing countries, with manufactures accounting for about 57 per cent by 1989. Changes in manufactured goods exports reflect changes in a country's industrial capacity – inevitable as a country industrialises. Developing countries have dramatically increased their share of global manufactured exports, especially the South and South-East Asian economies, which now account for over 75 per cent of the total manufactured exports from developing countries. But trade in manufactured goods is still dominated by the developed economies. Many developing economies, especially the poorest forty-seven, remain heavily dependent on primary commodity exports and markets. These commodities are often depressed and in unstable markets. Nevertheless, Nixson concludes that 'the empirical evidence on export-earning instability is inconclusive and does not suggest that developing countries as a group suffer extreme instability as a result of their dependence on primary products' (1996: 67).

Yet there is no doubt that the main problem usually highlighted in this kind of analysis is indeed the economic instability caused by overdependence on one or two raw material exports. The purchasing power of such exports may fluctuate quite dramatically, especially where the manufacture of synthetics or substitutes is common and where demand patterns are continually changing. Then, too, many other technological developments are likely to work against the developing countries simply because they reduce the possibility of advance towards industrialisation by transforming exhaustible material assets into valuable human-created wealth; the whole tendency of technological advance is to make material resources more homogeneous, both in quality and in their distribution, and to reduce the potential or actual value of high-quality natural resources that were once essential for industry. Again, increased agricultural productivity in the developed industrial countries has usually been accompanied by measures to protect those producers from adverse price effects arising from competition from low-income developing countries. Similarly, the protection of other domestic producers, such as oil and mineral producers, denies low-cost producers in the developing countries ready access to existing industrial markets. These fluctuations affect export earnings, balance of payments and customs duties.

The problem, then, is that low-income developing countries today are typically faced with fluctuating and often declining world markets for their traditional raw materials. The pattern of world trade and the terms of trade appear to favour the industrial countries at the expense of low-income countries.

That, at least, is the conventional wisdom, but much of it is now outdated and has less validity than it did. While it is still true that most developing countries export primary products (agricultural and mineral products) to the industrial economies in return for manufactured goods, this structure has changed significantly over the past thirty years or so. The developed market economies still dominate, but they account for declining proportions of developing countries' exports and imports. Trade between developing countries has expanded, but still makes up only a third of the total.

Changing patterns have also occurred in the over-dependence of some countries on a small number of exports – for example, bauxite and alumina from Jamaica, which account for 52 per cent of Jamaica's exports. But here again, changes are taking place in many developing countries. In Bangladesh, for example, whereas jute products accounted for 90 per cent of exports in 1972, by 1991 the share of jute had fallen to 18 per cent; ready-made garments accounted for over 50 per cent of the total, and a different use had been found for the product (Box 7.2).

Box 7.2

A future for jute in Bangladesh?

Bangladesh, when it was East Bengal in the days of the British Raj, was a colony devoted to the production of a single export crop: jute. The high rainfall and humid climate were perfect for growing the fibrous crop, for which there was a steady demand for use in the carpet and sacking industries, and later for the manufacture of linoleum. Yet the contribution jute made to the economy was less than it might have been, because all the manufacturing processes which add value to the raw material were carried out in Britain. The town of Dundee in Scotland became the 'jute capital' of the world.

In the twentieth century, however, the discovery of synthetic fibres such as polyethylene virtually destroyed the jute industry. Between 1970 and 1982 about 200 000 hectares of the Bangladesh countryside were taken out of jute production. Although Bangladesh still accounts for over 70 per cent of world production, farmers are reluctant to grow a crop when they are not guaranteed a good price for it.

Recently, however, new methods have been developed of reducing green jute to a pulp which can be made into paper. This was first developed in local factories, but India and South Korea are among those countries expressing an interest in paper made from jute pulp. There are already four large pulp mills in Bangladesh and the government is exploring the possibility of buying jute direct from farmers.

In a village some 50 km from Dhaka, seven women, with the technical assistance of the Intermediate Technology Development Group (ITDG) and economic assistance from a local NGO called the Socio-Economic Development Society, have set up a small factory. They produce handmade stationery and cards, and control the entire process, from the chemical preparation of the jute fibre, through machining the fibre into pulp, and eventually putting block-printed designs on the finished products.

Using jute in paper production on a large scale could reduce the number of trees which have to be cut down. At a time when international concern for the environment is becoming a powerful force, jute as a biodegradable and readily renewable resource may once again become the 'golden fibre'.

Source: Monan (1995: 46–7)

In East Asia, too, trading patterns have changed very markedly. The growth of trade has been dramatic and was largely responsible for the region's prosperity compared with other parts of the developing world. In so far as the negative and conventional points about trade and development summarised above are still valid today, they are most true of sub-Saharan Africa, where the growth of international trade has been very limited and sometimes negative (Figure 7.1). From 1982 to 1988 the growth of export volumes from East Asia (even excluding Japan) was just over 12 per cent, a rate double that of South Asia, three times that of the Middle East, North Africa and Latin America, and six times higher than in sub-Saharan Africa. At the same time, the growth of real GNP in East Asia and South-East Asia (excluding Japan) was twice that of South Asia, four times that of the Middle East, North Africa, Latin America and the Caribbean, and eight times that of sub-Saharan Africa.

World trading regions

Although the international trading movement of goods has a very long history, the scale of traded goods now moving beyond national boundaries within and between different regions has increased markedly with the rapid and recent growth of global linkages. At the same time, it is clear that these international trade flows are being increasingly

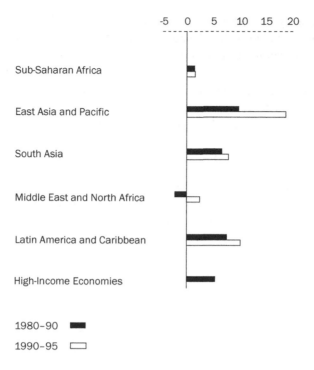

Figure 7.1 *Estimated percentage annual growth in real exports, 1980–90 and 1990–5*
Source: World Bank (1997b)

concentrated around a few supra-regions. These two rather conflicting trends have generated a debate – sometimes heated – over the role of regions in world trade.

Central to this discussion is the argument over whether trading regions (regionalism) are desirable. Those who support free trade (multilateralism) fear that the world economy could disintegrate into isolated trading blocs if the formation of regions is allowed to develop too far. This point of view is supported by those who argue strongly against the formation of regions created for trading purposes. Others, however, argue that regional trends in international trade are natural and that regions need not damage the development of free trade. After all, the impact of multinational corporations (MNCs) and their global–regional networks has inevitably had the effect of making regions more cosmopolitan and outward-looking. According to some authors, the argument over trade blocs or regions is unnecessary because research shows that regionalisation does not increase bloc traits; indeed, the truth is exactly the opposite. All trade regions or blocs have tended to become more global in scale. The conclusion is that regional cosmopolitanism thereby reinforces global interdependencies and realigns relationships between industrialised and developing countries away from those based on colonial ties or historical links. Regionalisation and multilateralism are complementary, not contradictory. Multilateralism and regionalism should not be viewed as distinct entities, representing entirely contrary and antithetical approaches to international trade. In the West Pacific Rim, multilateral, bilateral and regional approaches are all in evidence.

But all this depends to some extent on the region under consideration. In the West Pacific Rim the only grouping that approaches a trading region is ASEAN. But this is a relatively loose arrangement – an attempt to set up some form of integrating structure – which cannot reasonably be called a trading bloc. In the West Pacific Rim the priority of multilateralism over regionalism is marked and seemingly persistent. The reasons why there is no substantial regional trading bloc in the region are many, but two are worth identifying:

1 The centrifugal forces at work in the region – a very heterogeneous population and a physically immensely fragmented area – are much greater than in Europe or North America.
2 The dominant position of Japan – at least until now – has worked against regionalism in that the Japanese do not regard regionalism as the only or the best way to achieve global mulilateralism. As a result, the West Pacific Rim does not yet have any regional trading bloc, though it does have subregional groupings of an often loose and *ad hoc* kind.

An example of a similar kind of grouping is the Commonwealth of Independent States (CIS) – the former Soviet Union territory. As yet this is certainly not a trading bloc, but apparently there is every intention for the CIS to move towards the provision of some form of economic union. Then there is a true trade bloc in Latin America called Mercosur, formed between Brazil, Argentina, Paraguay and Uruguay. Recently Brazil's fallen currency has produced strains in Mercosur and calls for temporary limits to the bloc's free flow of goods. Even a large country such as Brazil knows that, in a trade world dominated by the United States, the EU and Japan, it would risk being disregarded on its own; and so would the other three member states.

Apart from the European Union, two of the most important trade blocs are NAFTA (the North America Free Trade Association), comprising the

United States, Canada and Mexico; and LAFTA (the Latin America Free Trade Association), comprising Brazil, Argentina, Mexico, Chile, Paraguay, Peru, Uruguay, Ecuador, Colombia, Venezuela and Bolivia.

As for APEC (the Asia-Pacific Economic Cooperation Forum), this is different from ASEAN, with its nine member states, or NAFTA, with its three signatories. APEC's members seem to form too disparate a group to work together effectively on detailed trade and investment issues. And the recent addition of three new members – Russia, Vietnam and Peru – makes the organisation seem even odder. But it could have a role in promoting good policies in global outlets such as the WTO or in providing individual governments with a platform for developing new ideas and projects.

Protectionism

Protectionism is the antithesis of trade liberalisation and free trade. It is generally accepted as undesirable and damaging to world trade and development. Tariff barriers may be erected to protect infant industries for a time, and such acts may be justified for short periods. But the use of traffic as a weapon in world trade, to attempt to damage another country's trade and prosperity for whatever reason, can easily develop into tit-for-tat actions that can lead to trade wars or worse.

But protectionism is now on the rise in a new guise – anti-dumping – and cases are now multiplying around the world (Table 7.1). Instead of raising import tariffs or cutting quotas, countries are slapping

Table 7.1 Anti-dumping cases, 1997

Main users		Main targets	
Australia	42	China	31
European Union	41	South Korea	16
South Africa	23	Taiwan	16
United States	16	United States	15
Argentina	15	Germany	14
South Korea	15	Japan	12
Canada	14	Indonesia	19
India	13	India	7
Brazil	11	Britain	6

Source: WTO, *Journal of World Trade*, reproduced in *The Economist*, 7 November 1998 'anti-dumping' duties on imports they claim are too cheap. The most prominent anti-dumpers are in the United States: in particular, the American steel industry is targeting imports from Japan, Brazil and Russia. South Korean memory-chip makers have been hit with heavy duties and their Taiwanese counterparts are now under fire. Many other anti-dumping cases are pending in industries such as machine tools, textiles and clothing. Canadian steel-makers are now following the US lead, and the EU is preparing anti-dumping measures against Japan, Taiwan and China.

Perhaps the most worrying trend in all this is that developing countries are now beginning to retaliate with anti-dumping actions of their own. Mexico, Argentina and Brazil have launched cases against the United States and others, as has South Africa. Asian countries are also hitting back: China, South Korea, India and Thailand are all taking anti-dumping measures.

Anti-dumping is difficult to control because World Trade Organization (WTO) rules allow it. WTO rules allow countries to impose anti-dumping duties on foreign goods that are being sold more cheaply than at home, or below the cost of production, when domestic producers can show that they are being harmed. On the other hand, it is possible to argue that countries which resort to anti-dumping measures have as their real motive, not easing the adjustment to freer trade, but bringing back protection by the back door. Certainly, developing countries have long criticised anti-dumping policy as disguised protectionism, arguing that it has allowed industrial economies to discriminate against their own exports. On the other hand, a growing number of developing countries have recently passed anti-dumping laws and are increasingly applying them to exports from rich trading partners.

The World Trade Organization

At the moment (mid-1999) the WTO is in some trouble, and not just because it cannot decide on a Director. It is also caught in a tug-of-war between the United States and Europe over bananas and hormone-treated beef. Little has gone right with the WTO recently in a period of weak world economies. Growth in world trade slowed to only 3.5 per cent in 1998 after being 10 per cent in 1997. When things go wrong, foreigners are a convenient scapegoat, justifying a surge of anti-dumping cases against 'unfairly' cheap imports. Even the United States is now moving

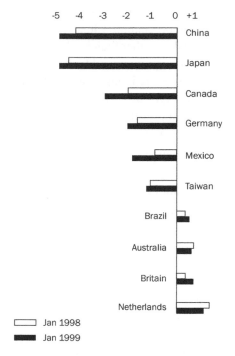

-5 -4 -3 -2 -1 0 +1

China
Japan
Canada
Germany
Mexico
Taiwan
Brazil
Australia
Britain
Netherlands

☐ Jan 1998
■ Jan 1999

Figure 7.2 *US monthly trade balances with selected countries, January 1998 and January 1999 ($bn)*
Source: World Bank (1999)

towards protectionism as a result of its own high trade deficit, problems with Japanese steel and much else (Figure 7.2). Trade relations with Europe are also bad. If Russia's economy slows still further, the mood could turn nasty. There is no doubt that the WTO is needed: its member countries have committed themselves to resolve their trade quarrels by asking an impartial panel of WTO experts to arbitrate, rather than by launching bilateral trade wars. And no member country can veto an unfavourable ruling.

The EU's 'misbehaviour' over bananas, beef and genetically modified foods reinforces the feeling in many developing countries that the WTO has one law for the rich and another for the poor. It is partly for this reason that some developing countries object to the fact that the WTO's new Director comes from New Zealand rather than from Thailand.

The WTO certainly needs to reform itself. It must become more open and transparent. And it is increasingly being asked to adjudicate on sensitive issues such as food safety and environmental protection that have become the stuff of trade disputes in a globalising economy.

China and the World Trade Organization

US objections to China's joining the WTO have been based largely on the issues of human rights, intellectual property, China's alleged stealing of US nuclear secrets, and the US bilateral trade deficit with China.

China is the largest trading nation still outside the multilateral system, with trade last year (1998) worth over $300 billion. Until China joins – which should happen imminently – the WTO will not be worthy of the name. If China were to join the WTO, that action would help provide greater access to China's markets. Exporters would also suffer from less uncertainty, because trade with China, based on genuine Chinese reforms, would be bound by international rules, rather than by the whims of politicians in Beijing or Washington. And exporters could call on the WTO to adjudicate when they felt their rights were being violated. More broadly, a deal with China would be a success for the United States' policy of engaging with China. Asia's biggest power, and the tenth largest trading nation, would be agreeing, in trade at least, to live by the same rules as the rest of the world. In time, that could promote cooperative behaviour in other areas.

For China would also reap benefits from joining the WTO. It would open new markets for its exporters and boost foreign investment in China – investment which has recently been flagging. It would also encourage further pro-market reforms and protect against any backsliding; it would protect against unilateral bullying by the United States; and consumers would gain from cheaper imports and greater competition. All this would provide the basis for China to revive its weakening economic growth. China also wants to join before Taiwan, which is much better prepared. One problem is that the Chinese economy is still largely controlled by the state: prices are fixed, monopolies common and subsidies legion. Its laws are opaque and often applied arbitrarily. Even if China does away with import quotas and dismantles tariff barriers, its markets will not be fully free in any meaningful sense. China will probably be allowed to join the WTO only if it pledges to liberalise a broad range of sectors and to apply its laws in a transparent and impartial manner.

Whatever happens over China, the WTO has had some reasonable success in its first four years. It has enforced world trade law much more vigorously than did its predecessor, the General Agreement on Tariffs and Trade (GATT), and it has dealt with 163 disputes since it was set up in 1995. The United States has brought the most cases (53), closely followed by the EU (43 cases). Developing countries have used the system against each other and against rich countries. Most cases have been settled without the need for arbitration, and as yet no government has formally defied its rulings. The WTO's greatest strength is that its rulings cannot be blocked by guilty offenders, even ones as powerful as the United States or the EU. Thus the weak can gain justice from the mighty. In spite of the present problems faced by the organisation, business everywhere has benefited from a disinterested enforcement of agreed international rules.

Aid

According to the latest report of the Organisation for Economic Co-operation and Development (OECD) on development assistance, China is the biggest single recipient of foreign aid. But compared with the size of its economy, China receives relatively little help: official foreign aid of $2.04 billion in 1997 was worth only 0.12 per cent of China's GDP – one of the lowest ratios among the emerging economies (*The Economist*, 13 March 1999). Other countries rely much more heavily on foreign aid. In five countries aid flows are worth more than 10 per cent of GDP: in Nicaragua official aid flows reached almost 23 per cent of GDP in 1997, and in Mozambique they amounted to over 37 per cent. But most developing countries have seen foreign aid flows decline during the 1990s, as rich countries have faced fiscal pressures. Nevertheless, six developing countries – Brazil, Bolivia, Peru, Thailand, Uganda and Vietnam – received more foreign aid in 1997 than in 1994. On the other hand, there are needy countries where aid is not forthcoming, for a variety of reasons: Myanmar with its poor reputation for human rights is one such example.

Rich countries are giving less aid to poor countries than ever before. On average, twenty-one OECD countries donated 0.22 per cent of their GNP to development assistance in 1997, down from 0.24 per cent a year earlier. The UN target, which only Denmark, Norway, Sweden and the Netherlands meet, is for rich countries to give 0.7 per cent of their national income in aid. But total overseas aid dropped from $55.4 billion in 1996 to $48.3 billion in 1997. Aid from the G7 group of large, rich countries has dropped by about 30 per cent in real terms since 1992. The United States now donates less than 0.1 per cent of its GNP.

The case of Bangladesh illustrates very clearly the problems as well as the opportunities associated with foreign aid. In the 1980s Bangladesh had become highly dependent on foreign funding for large new projects, such as power stations, bridges and roads, and for major items of the domestic budget, including the importing of raw materials and spare parts. But despite steady increases from year to year in external aid, the poor were still getting poorer. Between 1975 and 1987 per capita income actually dropped from $135 to $128; life expectancy remained constant at 48 years; infant mortality remained at a high level of around 130 deaths per 1 000 live births; and landlessness increased annually.

The billions of dollars in aid poured into Bangladesh had little effect on poverty because most of the money was spent on heavily capitalised infrastructure projects. These may have some indirect effect on the lives of poor people, but what people really want is better health and education services, better village roads rather than more motorways, and credit facilities to help them improve agriculture and raise their incomes. Most large Western donors found these small-scale projects too difficult to measure and adminster.

The degree of dependence by Bangladesh on aid had political consequences. Aid donors, allegedly, behaved as if their 'generosity' gave them the right to intervene in the running of the country; they therefore attempted to influence government on a wide range of issues. In the past few years, however, Bangladesh's dependence on aid has diminished. Investment is still sluggish, but with a respectable balance of payments, low inflation, recent stable commodity prices and a domestic budget entirely financed from local resources, Bangladesh is moving in the right direction. By financing more and more of its own development and investment, it will be able to break free of the political strings attached to foreign aid.

As for Britain's bilateral aid, the Overseas Development Administration (ODA) spent £55 million in Bangladesh on behalf of the British taxpayer. Most of this (£45 million) was in the form of project aid, of which half went to government infrastructure programmes of gas and electricity schemes and bridges. The remainder went on human development projects in the fields of health and population, education, and women in development, and was channelled through both the NGOs and the formal sector. A further £5 million was spent on training Bangladeshi professionals – between 400 and 500 people. Most of them went to British institutions, but in future training will be increasingly sought in institutes in the region. The ODA is now placing more emphasis on assessment and monitoring of projects. A further change has been to increase funding to Bangladeshi NGOs working in both rural and urban areas, which are to receive about £10 million. These positive changes have come about largely as a result of dialogue with NGOs and advocacy groups, which in the past have been critical of the way in which bilateral aid was used.

In 1992 funding from multilateral sources amounted to 53 per cent of the total aid disbursed to Bangladesh, which was roughly £1.25 billion. Over £500 million was in the forms of loans, mainly from the World Bank and the Asian Development Bank (ADB). In late 1994 the ADB committed a further £800 million to be spent over a four-year period. These loans may be 'soft' in that the interest charged is lower than for commercial banks, but they still have to be repaid; and they are dependent on the introduction of stringent economic measures. The banks insist on liberalisation of markets and deregulation of the economy before a loan is agreed. These conditionalities can have serious effects on poor people.

It is claimed in Bangladesh that some aspects of conditionality have created social problems for the government. Thus the demand that the jute industry be restructured led to the laying off of thousands of workers, which resulted in strikes and protests by men who faced unemployment and poverty after years of working in the industry. While market forces can play a powerful role in stimulating economic growth, structural adjustment programmes (SAPs) must be balanced with a concern for equity, and protection of the basic rights of the poor. The current World Bank policy of deregulating labour markets will merely increase 'poverty in employment' as wages fall; and there is little evidence to suggest that it will actually lead to economic growth.

Another unintentional effect of large amounts of bilateral and multilateral aid is the widespread corruption which it can create. The effects seem to leave no one untouched: contractors, consultants, bureaucrats and politicians all have much to gain personally from the flow of aid. The control of aid projects gives immense power to politicians and bureaucrats, and fortunes can be made by suppliers and contractors. The power and patronage reach down to deciding who gets the job of breaking the stones in the construction of rural roads. The result is an ever-widening gap between the minority elite and the impoverished 95 per cent of the population. The flow of aid may be adding to the problem of gross inequalities in Bangladesh society.

A further problem associated with increasing aid flows to a country is the assumption that an economy needs aid before it can make any progress in its development. But as the case of Vietnam shows (Figure 7.3), a great deal can come from knowledge dissemination sponsored by development agencies, even when unaccompanied by substantial financial transfers. Vietnam was plagued in the mid-1960s by hyperinflation, a huge fiscal deficit, poor incentives for production, and stagnant income per capita. The country began to reform in 1986, but because of its political estrangement from the West, it received no large-scale financial assistance. Vietnam did, however, receive a significant amount of technical assistance and policy advice, financed by the Nordic countries and the United Nations Development Programme (UNDP). Both the World Bank and the International Monetary Fund (IMF) were active in delivering this assistance and advice. Only after a marked policy improvement between 1988 and 1992 did significant amounts of financial assistance begin to flow into the country in a sustained way. But by then a sharp improvement in economic performance had already taken place: income per capita was growing strongly, and inflation had fallen off dramatically, from over 400 per cent in 1988 to 32 per cent in 1992. The important lesson from Vietnam's experience is that donor agencies can help with policy reform and institutional development before providing large amounts of money (World Bank 1999).

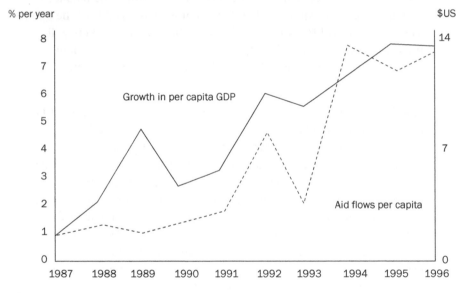

Figure 7.3 *Aid flows and GDP per capita in Vietnam*
Source: After World Bank (1999)

In view of the comments made about the problems associated with aid-giving, it is hardly surprising that there are clear differences in perspective about the value of foreign aid in development. Some writers regard it as a means of creating and perpetuating dependency – as imperialism. And it is true that much foreign aid does not match up to the expectations placed upon it: (a) that it should be uncommercial from the point of view of the donor; and (b) that it should carry interest and repayment terms which are less stringent than those prevailing in the commercial world (Elliott 1999: 74). The point has already been made that foreign aid can easily worsen rather than improve developing economies in one way or another. On the other hand, there is a clear and growing recognition that foreign aid, properly administered in the real interests of recipient countries, must be an essential element of a mutually beneficial partnership between developed and developing countries.

Summary

The discussion in this chapter centres on three related issues: regional groupings of states, world trade blocs and aid. The logic of grouping states together for mutual cooperation and development is not always matched by success in the way the individual states react to the strains and responsibilities of cooperation. The problems most states face are similar to those experienced in many states of the relatively successful European Union. As for the creation of world trade blocs, this issue raises the controversy over how best to improve and control world trade – by regionalism (i.e. through blocs) or by multilateralism. The World Trade Organization (WTO) will, it is hoped, be more effective in stimulating, directing and controlling world trade than was its predecessor, the GATT. As for aid, the case of Bangladesh is used to illustrate the problems as well as the opportunities associated with foreign aid.

Questions for discussion

1 With reference to specific cases, assess the arguments for and against the creation of regional groupings of states as the frameworks for development.
2 Contrast the effectiveness of regionalism and multilateralism in world trade.
3 How far do you agree with the view that most foreign aid is a form of imperialism?

Further reading

Auty, R. (1995) *Patterns of Development* London: Arnold.

Gibb, R. and Michalak, W. (eds) (1994) *Continental Trading Blocs: The Growth of Regionalism in the World Economy* London: Wiley.

Helleiner, G.K. (1992) 'The IMF, the World Bank and Africa's adjustment and external debt problems: an official view' *World Development* 15, supplement.

Tarrant, J.R. (1985) 'A review of international food trade' *Progress in Human Geography* 9: 235–54.

8 Globalisation

- The process of globalisation
- The debate on globalisation
- Implications for developing countries: multinational corporations, debt, sustainable development
- Conclusions

Chapter 7 discusses trade and aid, which have for some time now been important indicators of increasing global interdependence. In recent years, however, the debate on global interdependence has gained greater momentum within a more general perceived process known as globalisation.

The process of globalisation

The term 'globalisation' has been defined in many different ways. It has been described as referring to the process whereby the world's economies, societies and cultures are becoming ever more closely intertwined (Giddens 1999). Earlier, it was given rather greater precision by Dunning, who defines globalisation as the process giving rise to

> an economy in which there is close economic interdependence among and between the leading nations in trade, investment, and cooperative commercial relationships, and in which there are relatively few artificial restrictions on the cross-border moverment of people, assets, goods or services.
>
> (1993: 124)

As Knox (1995) points out, however, globalisation is not a new phenomenon. The key role in international finance played by just a few key cities, the growth of many international agencies, standardised systems of international time and communication, and a shared acceptance of such universal values as human rights: all these have their origins in the nineteenth century or earlier. What has been distinctive about globalisation in the late twentieth century – and one reason why the term has become so fashionable today in the literature – is that it has recently acquired some new elements. First, there has been

> a decisive shift in the proportion of the world's economic activity that is transnational in scope. At the same time, there has been a decisive shift in the nature and organisation of transnational activity, with international trade in raw materials and manufactured goods being eclipsed by flows of goods, capital, and information that take place within and between transnational conglomerate corporations. A third distinctive feature, interdependent with the first two, is the articulation of new world views and cultural sensibilities: notably the ecological concern with global resources and environments and the postmodern condition of pluralistic, multicultural, non-hierarchical, and de-centred world society. . . . Much of this change has been transacted and mediated through world cities, the nodal points of the multiplicity of linkages and interconnections that sustain the contemporary world economy.
>
> (Knox 1995: 233–4)

Although globalisation is by definition a world phenomenon, and involves the flow of information and knowledge around the world, this does not mean that it is universal or even in its impact. As Dunning points out in the definition quoted above, it concerns especially the 'leading nations', and has relatively little effect on developing economies in terms of its impact on global economic disparities (Figure 8.1). Certainly as yet it does not imply any evening out of the roles of the developed and developing economies, and a number of writers refer to the transfer upwards of real power to the global context. Global connectedness operates to some extent above the developing countries as developed economies play the lead role.

Nor does globalisation imply any diminution or disappearance of geographical scale. Globalisation does not in any sense make other geographical scales disappear. Yeung (1998) points out that the globalisation of economic activities and transnational corporations (TNCs) has encouraged some writers to suggest that we are entering what he calls a 'borderless' world – a world which must lead to the end of the concept of the state and which regards capital as 'placeless'. Some even suggest that it marks the end of geography. Yeung argues against

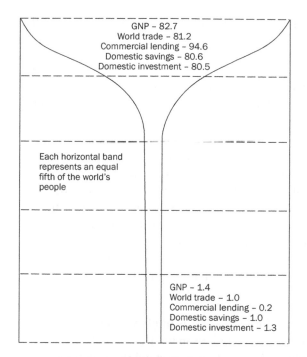

GNP – 82.7
World trade – 81.2
Commercial lending – 94.6
Domestic savings – 80.6
Domestic investment – 80.5

Each horizontal band represents an equal fifth of the world's people

GNP – 1.4
World trade – 1.0
Commercial lending – 0.2
Domestic savings – 1.0
Domestic investment – 1.3

Figure 8.1 *Global economic disparities*
Source: UNDP (1992) (data 1989)

this notion, noting that despite the accelerated processes of globalisation, national boundaries still matter in the decision-making and global reach of capital. Globalisation need not mean the disappearance of national and regional dimensions. In this sense, and to some extent, the issue of globalisation is remniscent of the arguments over the European Union: does it involve or imply abandoning national ties and embracing supranational alliances? Does globalisation place impossible pressures on national governments?

In recent discussion on the globalisation of the economy there has inevitably been some focus on what has been called 'the crisis of capitalism'. Certainly as far as the developing economies are concerned, the broad global economic picture looks unpromising. In 1998 economic growth in developing countries was only 2 per cent, roughly half its 1997 level. Private capital flows to emerging markets have all but dried up and will not quickly recover. Some thirty-six countries, accounting for 40 per cent of the developing world's GDP, probably suffered negative per capita growth in 1998. All across East Asia and Latin America the social fabric continues to tear as poverty increases, with unforeseen economic and political consequences. In the five largest emerging markets development difficulties are especially worrying. Brazil is seriously indebted, Indonesia is on the verge of social revolution, Russia's economy is still moving downwards, and China's economic reforms are under great stress as exports slow dramatically, growth rates fall well short of their targets, and the regional banks default on foreign debts. India's progress is held back by internal political paralysis. And all this is happening against the backdrop of a 50 per cent decline in the growth of global trade from 1997 to 1998 and amid an inevitable slowdown in the United States and

Europe, the only two regions of the world where growth rates are healthy (Garten 1999: 77).

In this context it is feared that in the periphery – the developing countries – economic failure is in danger of becoming so intense that individual countries will begin to opt out of the global capitalist system altogether. Soros, in his book *The Crisis of Global Capitalism* (1998), argues that the choice is whether we will or can regulate global financial markets internationally or simply leave it to each individual state to protect its own interests as best it can. According to Soros, the latter course will lead to the breakdown of the gigantic circulatory system which goes under the name of global capitalism. Such a breakdown can, he believes, be prevented only by the intervention of the international authorities. Soros also denounces the 'wrecking ball' of international capital, predicting that 'the political developments triggered by the financial crisis may eventually sweep away the global capitalist cystem itself' (*Foreign Affairs* March/April 1999, 78, 2: 124–31).

Whatever the causes or solutions of the crisis of global capitalism, this crisis has eroded the faith of many nations in modern capitalism itself. A quiet backlash is gradually building up against unbridled economic liberalisation and the chaos it seems to have unleashed. Leftwich contends that democratic politics cannot promote development; development requires 'a strong developmental state that can pursue "market friendly" policies in such a way that even poor households are empowered to participate in development' (1993: 437). The activities of the World Bank and the severe conditions imposed by the IMF also affect this issue.

On the other hand, many countries have had to face up to the turmoil of moving from a political system in which power was confined to a relatively restricted elite to a more inclusive democracy. The leftish views and predictions of a decade or so ago – that, for instance, the poor will eventually rise up and demand global justice, or that the English-speaking global culture is threatened everywhere by a fundamentalist and anti-Western backlash – are no longer accepted. By combining capitalism with advancing technology, political democratisation and a globalised culture, Davos Man – referring to members of the World Economic Forum, who meet annually in Davos, Switzerland – is creating a much firmer framework than before. He has also benefited from the tide of ideological moderation – left of centre and supporting welfare capitalism (Kaletsky 1999) – which now dominates the global political scene.

The debate on globalisation

Those arguing the case *for* globalisation believe that it has great potential power for good. This new, fashionable thesis was presented and discussed by Anthony Giddens in the 1999 Reith Lectures (a series of lectures given annually on behalf of the British Broadcasting Corporation): we live, we are told, in an essentially new world. It is one controlled by the global economy, society, culture and citizenship (Giddens 1999).

In economic terms, globalisation involves more trade, which is presumed to be better for all parties concerned, for the goal of self-sufficiency makes no sense in a closely integrated world economy with free buying and selling.

A number of cases illustrating this point have been made recently. Gwynne (1999) has examined the impact of globalisation and its associated commodity chains in the fruit-exporting regions in Chile. And Barrett *et al.* (1999) have discussed globalisation and the changing networks of food supply with special reference to the importation of fresh horticultural produce from Kenya into the United Kingdom. They point out that the trade in fresh horticultural products has become increasingly global, and demonstrate the complexity and interdependence of the various scales, from local to global. The wholesale network is based on

> international ties, often based on kinship, and dependent for its trading success on trust, flexibility and mutual agreement. The supermarket chains rely on various types of contract and agreements, in a descending power sequence of supermarket, importer and supplier. Both of these marketing chains must take account of the international and national regulatory framework which requires that produce meets quality and food safety standards . . . the increasingly global market for horticultural produce is thus drawing Kenyan farmers, operating at all scales, into the network of supply.
>
> (Barrett *et al.* 1999: 173)

Again, globalisation encourages an open-door policy with regard to multinational corporations (MNCs), which have largely beneficial effects on the economies of developing countries. The very largest MNCs are now responsible for four-fifths of the foreign direct investment and half of world trade, and sit at the centres of webs of relationships with suppliers and service companies throughout the world.

Globalisation also improves world knowledge and communication (Table 8.1). It can democratise and empower. It will increasingly connect and inform even those relatively cut off from the rest of the world in developing countries; and information is power (Janet Bush, *The Times*, 3 February 1999).

Table 8.1 *Selected indicators of information and telecommunications penetration by country income level*

Group of economies	Telephone main lines per 1 000 people	Personal computers per 1 000 people	Internet users per 1 000 people
Low-income	26	2	0
Lower-middle-income	95	10	1
Upper-middle-income	130	24	4
Newly industrialising countries	448	115	13
High-income	546	199	111

Source: World Bank (1999)

Politically, globalisation can be regarded as an opportunity and a positive force. It can act as a counterbalance to the intrusive powers of the state, and opens up new opportunities for nations to work together. At the same time, globalisation does not imply any erosion of a democratic state's sovereignty, nor does it inhibit full consideration of the regional dimension.

The arguments ranged *against* globalisation are equally numerous and powerfully expressed. Thus globalisation – or globality – is 'a graceless system that renders people surplus and abandons them if they cannot compete in the global economy' (D. Selbourne, *The Times*, 2 March 1999, p. 18). Globalisation in the sense of travel is the preserve of a tiny elite. Advocates of the dependency school have always argued that integration with global capitalism – which is what globalisation involves – would hinder development in the peripheral countries. Furthermore, just as Fukuyama's 'end of the historical process' after the collapse of communism has been invalidated by subsequent events, so 'globalisation' will turn out to have been another chimera.

According to several writers, too, the globalisation thesis expresses a strong, even apocalyptic, death-wish for the nation-state and the moral order. Globalisation implies that the nation-state is out of date, having been superseded by global forces that dwarf it and which it cannot control. Again, the thesis of globalisation provides a cover, or legitimation, for the failures of individual states to tackle their manifold social, economic, ethical and environmental problems. Furthermore, the thesis of globalisation is no more than a particular, and limited, construction of reality. A global vision is in fact myopic: it sees the wood rather than the trees. It is also a truism, for the world is by definition global, and can be no other. There is nothing new about it today, except in degree. National policy decisions, cultural traditions and social conditions remain the important variables in determining a state's economic fortunes. In common with its predecessor, the 'end of history', it promises much as an explanatory notion in our dark times, yet in the end provides no real illumination at all.

The nation-state, then, is still the most important place where democracy, accountability and power coincide. This is the antithesis of globalisation, which 'implies abandoning nationalities and embracing supranational alliances' (Sassen 1999: 81). Globalisation is frequently referred to by its detractors as the latest apocalypse galloping the globe. Rather than ushering in a golden age of open markets and liberal democracy, globalisation is a malign and volatile voice.

Some implications of globalisation for developing countries

The major and underlying problem of globalisation for developing countries is that any downturn in the global economy is bound to affect developing economies significantly more than developed economies. The World Economic Forum at Davos (February 1999) was dominated by an attempt to analyse the implications of globalisation. The theme of the forum was 'Responsible Globalisation: Managing the Impact of Globalisation'. The section that follows examines a few of the implications of this impact for developing economies.

Transnational corporations

There are strikingly opposing views about the role of TNCs or MNCs in developing countries. It is certainly a matter of some importance:

In the early 1990s there were 37 000 TNCs with over 170 000 foreign affiliates (UN 1993). The total stock of foreign direct investment (FDI)

accounted for by this universe of corporations was in excess of $2 trillion
in 1992, and totalled over $5.5 trillion in worldwide sales.

(Kaletsky 1999)

Moreover, TNCs' capital is highly concentrated. Ninety per cent of
TNCs are in advanced capitalist states, and the 'triad of Japan, North
America and Western Europe produced 70 per cent of global foreign
investment inflows and 96 per cent of global outflows in 1991'
(Johnston *et al.* 1998: 13).

Dasgupta (1998) examines the continuing rise in the power and role of
TNCs and comments on their poor reputation, caused largely by their
alleged exploitative treatment of workers. It is true that multinationals
have now greatly improved their working conditions in most developing
countries, and scrutiny of their own plants has improved a great deal; but
they are still vulnerable to public relations disasters. Codes of conduct
which set out the standards a multinational expects of its factories and
contractors have evolved from vague promises into detailed rules. The
best codes now tend to be monitored by outside auditors, though there is
a problem in gathering credible evidence for bad practice. Yesterday the
criticism was child labour and prisoners: today it is long hours and low
pay. In future, activists may have stronger arguments, but whatever
scandals they uncover, real improvements will come only if there is
clearer thinking and better research (Box 8.1).

It is easy to criticise the activities of many MNCs, and there is no doubt
that improvements are still necessary. But to move from that to a blanket
disapproval of all MNCs and to argue that they should not be allowed
into developing countries would do no good at all. Judgement must be
balanced. Many MNCs have had positive impacts on the economies and
societies of poor countries. It is true that they take advantage of low-
wage labour; but MNCs also play a role in the development of a new
international division of labour (NIDL) in Asia and Latin America. It
should also be remembered that MNCs favour

> developed countries primarily, and South America and South and East
> Asia secondarily. Within the global periphery, multinational corporations
> invest heavily in locations possessing an Export Processing Zone (EPZ),
> places where the main lures include not only relatively inexpensive
> wage labour, but also land, electricity, and water at cheap rates, as
> well as government grants, tax breaks, tariff/duty reductions,
> lax pollution controls and less rigorous health and safety standards in the
> workplace.
>
> (Barff 1995, quoted in Dasgupta 1998: 60–1)

Box 8.1

Multinationals: Disney and Mattel toy firms in China

Reports have alleged that some workers making Disney products are forced to work for up to sixteen hours a day, seven days a week, and are paid almost no overtime. In a report on toys, four of the twelve plants singled out for criticism are subcontractors to Mattel, the world's largest toy maker. What have come to be known as 'sweatshop Barbie' assembly lines are accused of a range of abuses, ranging from long hours and low pay to heavy fines for workers' infringements.

Multinationals have often been linked to factories that are grim by First World standards. Yet the best-known multinationals have improved of late. Disney and Mattel have done more than most other firms to improve working conditions in their Asian plants. Both have codes of conduct; and the independent panel that Mattel has set up to monitor its factories is considered a model in the industry, even by activists. Social Accountability 8 000 is an independent certificate awarded by the Council on Economic Priorities, an interest group based in New York. Thanks to this and similar initiatives, auditors are doing a brisk business, as with PriceWaterhouseCoopers, which in 1998 conducted 1 500 inspections in Guangdong province alone.

Yet even the best corporate ethics programme will not end complaints about cheap labour. Multinationals make things in Asia because wages are meagre by Western standards: from $3 a day to as little as 30 cents. Although most factories have no shortage of applicants, that is not good enough for those who want a 'fair living wage', defined as one that will support half the worker's family. This is hard to define. In expensive Guangdong, for instance, a worker's earnings would not support half a family, but many workers come from the Chinese interior where their earnings *would* support at least half their family.

Source: *The Economist*, 27 February 1999: 76–9

The TNCs have taken great care to alleviate the politician's biggest fear about their presence in a country: that they will act politically. Most TNCs have shown that they can be steadfast in a country's best interests and that they can be a stable source of capital and technology. Asian economies, as well as the companies themselves, have much to gain if TNCs are treated more fairly. A good start would be to scrap the rules and practices that have discriminated against foreign companies.

Plate 14 *Multinationals and anti-imperialism*

Photograph reproduced courtesy of Panos Pictures; © Paul Smith

Debt

According to recent figures from the World Bank, Brazll had the largest external debt at the end of 1997, owing almost $200 billion. But the burden of external debt depends less on its size than on a country's ability to service it. Asia's financial crisis slowed net flows of private credit to developing countries substantially in 1997. Yet their debt fell only slightly. Developing countries' foreign debt declined from 35.8 per cent of their combined GDP in 1996 to 34.9 per cent in 1997.

Early in 1999 the US President, Bill Clinton, called for poor countries to be released from some of their debts, recognising the fact – obvious for a long time – that several countries, especially in Africa, cannot repay those debts. Their occasional efforts to do so impoverish already destitute people and blight their hopes of economic take-off. Attempts have been made: the Paris Club of official creditors has offered steadily better terms to poor countries; commercial banks have become used to buying back their loans at deep discounts; and in 1996 the Heavily Indebted Poor Countries (HIPC) initiative was launched by the World Bank and the IMF. This broke a taboo on the structuring of debts to multilateral agencies and set up a framework for systematic debt reduction in return for overdue economic housekeeping. Some forty countries have now been classified as HIPCs. They owe $170 billion, less than half the debt owed by low-income countries as a whole. On average their present debts exceed their annual export earnings by more than fourfold (*The Economist*, 20 March 1999: 23).

Arguments commonly advanced against forgiving or cancelling debts include the notion that most countries are poor not because they are in debt, but because they are run by incompetent and corrupt governments. This is believed to be true of many governments, such as those of Cambodia, Laos and Vietnam. However, the decision to reduce the debts of the poorest developing countries has now been made; and in 1999 the G7 countries cancelled $70 billion worth of debts in the poorest economies.

Anticipating the next subsection of this chapter, we should be aware that the debt burden of many developing countries has two major implications for the prospects of sustainable development:

1 the need 'to increase short-term productivity puts pressure on countries to overexploit their natural resources' – thus in the 1980s

Brazil, the most heavily indebted country, suffered the greatest extent of deforestation;

2 the 'level of government austerity necessitated by debt servicing reduces a government's capacity to deal with environmental protection and rehabilitation: money diverted to servicing debt is unavailable for environmental management (or indeed, wider programmes of poverty alleviation)' (Elliott 1999: 86).

Sustainable development

Of particular relevance to the dangers of the possible collapse of the global capitalist system and a retreat into national independence in financial matters is the whole question of sustainable development. Increasing global environmental degradation is undoubtedly a concomitant of global economic development, though as yet it is a greater problem in developed than in developing countries. The future, however, is likely to see a very different scenario. Sustainable development was defined by the Brundtland Commission (1987: 43) as 'development which meets the needs of the present without compromising the ability of future generations to meet their own needs'. Or, as Potter *et al.* (1999: 108) put it more precisely, 'the essence of sustainable development is the need to achieve an equilibrium between the world's basic resources and their continuing exploitation by a growing world population, so as not to jeopardise the resources for future generations'. Sustainable development was the driving force behind the 1992 Earth Summit (UNCED) in Rio de Janeiro.

The range of issues covered by the term 'sustainable development' includes almost everything discussed in the various chapters of this book on development geography, though the emphasis is quite distinctive – on, for instance, the environment, its pollution and resources; rural and urban livelihoods; and poverty. This vast field of sustainable development in the developing world has been admirably summarised and discussed by Elliott (1999). She demonstrates that 'unprecedented rates and degrees of environmental, economic and political change currently impact on and connect people and regions across the globe' (p. 176). The numerous examples and case studies presented in Elliott's book add up to a convincing statement on the significance of sustainable development as a central focus in the study of development geography.

Plate 15 *Digital communications being installed in Botswana, Africa*

Photograph reproduced courtesy of Panos Pictures; © Trygve Bølstad

Plate 16 A computer cartographer in Los Banos, Luzon, the Philippines

Photograph reproduced courtesy of Panos Pictures; © Heldur Netocny

Conclusions

The arguments for and against globalisation, however defined, will no doubt continue, especially as many of them seem to be based on misunderstandings and misinterpretations of one kind or another. But regarded primarily not as a new phenomenon, but simply as a new term to describe the dynamics of contemporary economic, social and political life, globalisation is a useful and important concept which must increasingly demand attention in development geography.

One might think that globalisation, in which all parts of the world can be so readily and easily reached, must result eventually in the convergence of the developed and developing worlds. The division of the world by scholars into these two worlds – one rich, the other poor – seems an increasingly anachronistic device. But it is becoming clear that access to knowledge and information does not guarantee greater access to equality of economic opportunity or to sufficient amounts of the factors of production necessary for rapid economic advance. As yet there is certainly no sign that globalisation will work to the benefit of developing countries. Indeed, there is some evidence that exactly the opposite is occurring – that developing countries are being by-passed by the increasing concentration of power within larger and larger organisations – 'global giants'. Many of these global giants are created by mergers, and almost all of them are associated with the major urban nodes in the advanced, industrialised countries of the developed world.

On the other hand, it is not impossible to conceive of a future in which the concentration of economic power in a few centres in the developed world changes quite

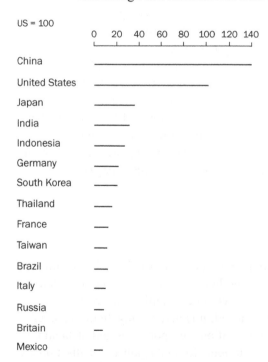

US = 100

| | 0 | 20 | 40 | 60 | 80 | 100 | 120 | 140 |

China
United States
Japan
India
Indonesia
Germany
South Korea
Thailand
France
Taiwan
Brazil
Italy
Russia
Britain
Mexico

Figure 8.2 *The fifteen largest economies in 2020?*

Note: Forecasts assume that countries continue to grow at regional rates projected by the World Bank. GDPs are at purchasing power parity (PPP)
Source: World Bank (1999)

radically. For example, we tend to assume that today's pattern of economies is almost cast in stone, whereas, as Figure 8.2 suggests, the pattern of economies by the year 2020 could be very different, with China the world's largest economy.

But whatever scenario turns out to be correct, the question still needs to be asked and considered: is it time to get rid of the distinction between the developed and developing worlds in our thinking and writing? As this short book has illustrated, geography is a very wide-ranging, eclectic, global and interdisciplinary subject. And yet it is both implied and stated in much of the literature on developing countries that a development geography which spans both the developed and developing worlds is impossible. Thus although Corbridge argues that developing countries can be understood only in relation to the developed countries, he goes on to conclude pragmatically: 'no wonder that no-one has yet written the Great Development Studies text' (1995: xi). Similarly Todaro, writing as an economist, rejects the view that the study of development economics is simply the same as that of other parts of the subject, except that it focuses specifically on the 'poor' regions of Asia, Africa and Latin America. He contends that development economics is a field of study 'that is rapidly evolving its own distinctive analytical, methodological identity . . . it is the economics of contemporary poor, undeveloped Third World nations with varying ideological orientations' (1997: 7). Whether these viewpoints are valid and whether convergence between the developed and developing worlds is already occurring are matters which demand much greater attention than is possible in these pages.

Summary

There is some difference of opinion about what exactly globalisation means, whether it in fact exists, whether it is really anything new, and what its implications are. But whatever one's opinion, this new, fashionable concept cannot be ignored; it is here to stay. In this chapter particular emphasis is laid on transnational corporations, aid, debt and sustainable development. The arguments for and against globalisation are summarised. Finally, some comments are made about the implications of current thinking on globalisation for development geography – in particular, on how long the distinction between the developed and developing worlds can be maintained in a truly global economy.

Questions for discussion

1 Critically examine the arguments for and against globalisation as a useful concept in development geography.
2 How might globalisation bring about convergence between the developed and developing worlds?
3 To what extent might globalisation assist or hinder success in reaching the goals set by the OECD and listed in the Epilogue to this book?

Further Reading

Berry, B.L.B., Conkling, E.C. and Ray, D.M. (1997) *The Global Economy and Transition* Hemel Hempstead: Prentice Hall.

Bryson, J., Henry, N., Keeble, D. and Martin, R. (eds) (1999) *The Economic Geography Reader: Producing and Consuming Global Capitalism* Chichester: Wiley.

Daniels, P.W. and Lever, W.F. (eds) (1996) *The Global Economy in Transition* London: Longman.

Dasgupta, B. (1998) *Structural Adjustment, Global Trade and the New Political Economy of Development* London: Zed.

Elliott, J.A. (1999) *An Introduction to Sustainable Development: The Developing World* 2nd edition, London: Routledge.

Giddens, A. (1999) *Runaway World: How Globalization is Reshaping Our Lives* London: Profile.

Johnston, R.J., Taylor, P.J. and Watts, M.J. (eds) (1998) *Geographies of Global Change: Remapping the World in the Late Twentieth Century* Oxford: Blackwell.

Epilogue

Are we justified in adopting a cautiously optimistic viewpoint about the future of development? In a book of this kind it is perhaps inevitable that so much attention is directed at the problems of development to the extent that we lose sight of all progress, however small, towards the goals of development. These goals are clear enough. As summarised by the Development Assistance Committee of the Organisation for Economic Co-operation and Development (OECD), they are:

- to reduce by half the proportion of people living in extreme poverty by 2015;
- to achieve universal primary education in all countries by 2015;
- to demonstrate progress toward gender equality and the empowerment of women by eliminating gender disparities in primary and secondary education by 2005;
- to reduce by two-thirds unfant and child mortality rates and by three-quarters maternal mortality rates by 2015;
- to provide access to reproductive health services for all individuals of appropriate age by no later than 2015;
- to implement national strategies for sustainable development by 2005 to ensure that the current loss of environmental resources is reversed globally and nationally by 2015.

While it is probable that almost everyone will agree with these aims, not everyone will agree about how they should be achieved; and this highlights once again the point reiterated throughout these pages – that there is no 'correct' development path, and that every interpretation of a problem and how to solve it depends on one's practical and theoretical perspective. In some quarters, at least, the ideological debate still rages.

However, it appears that there is a new, if somewhat reluctant, consensus on development emerging, leading to a more pragmatic approach to decisions about development strategies. As one writer has expressed it, 'capitalism at present, with all its strengths and weaknesses, remains the framework within which actions for sustainable development have to be taken . . . the system is sufficiently flexible to enable multiple views of environment and development and diverse solutions in practice' (Elliott 1999: 178).

Bibliography

Adams, W.M. (1990) *Green Development: Environment and Sustainability in the Third World* London: Routledge.

Aikman, D. (1986) *Pacific Rim: Area of Change, Area of Opportunity* Boston: Little Brown.

Amin, S. (1976) *Imperialism and Unequal Development* Hassocks, West Sussex: Harvester.

Anderson, A. (1990) (ed.) *Alternatives to Deforestation: Steps towards Sustainable Use of the Amazon Rainforest* New York: Columbia University Press.

Apter, D. (1987) *Rethinking Development: Modernization, Dependency and Postmodern Politics* Newbury Park: Sage.

Auty, R. (1993) *Sustaining Development in Mineral Economies: The Resource-Curse Thesis* London: Routledge.

Auty, R (1995) *Patterns of Development: Resource Endowment, Development Policy and Economic Growth* London: Arnold.

Baer, W., Miles, W.R. and Moran, A.B. (1999) The end of the Asian myth: why were the experts fooled? *World Development* 27, 10: 1735–47.

Bairoch, P. (1975) *The Economic Development of the Third World since 1900* London: Methuen.

Baran, P. (1973) *The Political Economy of Growth* Harmondsworth: Penguin.

Barrett, H.R., Ilbery, B.W., Browne, A.W. and Binns, T. (1999) Globalisation and the changing networks of food supply: the importation of fresh horticultural produce from Kenya into the UK' *Transactions of the Institute of British Geographers* New Series 24, 2: 159–74.

Bauer, P. (1972) *Dissent on Development* London: Weidenfeld and Nicolson.

Bauer, P.T. and Yamey B. (1957) *Economic Development in Underdeveloped Countries* Cambridge: Cambridge University Press.

Becker, B.K. and Egler, C.A.G. (1992) *Brazil, a New Regional Power in the World Economy* Cambridge: Cambridge University Press.

Bernstein, H., Crow, B. and Johnson, H. (eds) (1992) *Rural Livelihoods: Crises and Responses* Oxford: Oxford University Press.

Berry, B.L.B., Conkling, E.C. and Ray, D.M. (1997) *The Global Economy and Transition* Hemel Hempstead: Prentice Hall.

Bhagwati, J. (1993) *India in Transition: Freeing the Economy* Oxford: Oxford University Press.

Binns, T. (ed.) (1995) *People and Environment in Africa* Chichester: Wiley.

Binns, T. (1994) *Tropical Africa* London: Routledge.

Blaut, J. (1993) *The Colonizers' Model of the World* London: Guilford.

Booth, D. (1985) Marxism and development sociology: interpreting the impasse' *World Development* 13: 761–87.

Bose, C.E. and Acosta-Belem, E. (1995) *Women in the Latin American Development Process* Philadelphia, PA: Temple University Press.

Boserup, E. (1993) *The Conditions of Agricultural Growth* London: Earthscan.

Brandt, W. (1980) *North–South: A Programme for Survival* London: Pan.

Brohman, J. (1996) *Popular Development: Rethinking the Theory and Practice of Development* Oxford: Blackwell.

Brookfield, H. (1975) *Interdependent Development* London: Methuen.

Brundtland, G.H. (1987) *Our Common Future: The World Commission on Environment and Development (WCED)* Oxford: Oxford University Press.

Bryson, J., Henry, N., Keeble, D. and Martin, R. (eds) (1999) *The Economic Geography Reader* Chichester: Wiley.

Chambers, R. (1983) *Rural Development: Putting the Last First* London: Longman.

Chambers, R. (1997) *Whose Reality Counts?* London: Intermediate Technology Publications.

Chandra, R. (1992) *Industrialization and Development in the Third World* London: Routledge.

Chant, S. (ed.) (1992) *Gender and Migration in Developing Countries* London: Belhaven.

Chant, S. (1996) *Gender, Uneven Development and Housing*, New York: United Nations Development Programme. Anderson, J., Brook, C. and Cochrane, A. (eds) *A Global World*, Oxford: Oxford University Press and Open University.

Clapp, R.A. (1998) Waiting for the forest law: resource-led development and environmental politics in Chile' *Latin American Research Review* 33, 2: 19–28.

Clark, D. (1998) Interdependent urbanisation in an urban world: an historical overview' *Geographical Journal* 164, 1: 85–95.

Cohen, A. (1971) *Custom and Politics in Urban Africa*, Berkeley: University of California Press.

Cohen, A. (1976) *Two-Dimensional Man*, Berkeley: University of California Press.

Collins, C.O. (1995) Refugee resettlement in Belize' *Geographical Review* 85, 1: 20–30.

Copper, J.F. (1990) *Taiwan: Nation-State or Province?* Boulder, CO: Westview.

Corbridge, S. (1992) Third World development' *Progress in Human Geography* 16, 54: 584–95.

Corbridge, S. (ed.) (1995) *Development Studies: A Reader* London: Arnold.

Cowan, M.P. and Shenton, R.W. (1996) *Doctrines of Development* London: Routledge.

Craig, M. (1998) *Cultural Geography* London: Routledge.

Crook, N. (1997) *Principles of Population and Development* Oxford: Oxford University Press.

Crush, J. (1993) Postcoloniality, decolonization, geography' in Godlewska, A. and Smith, N. (eds) *Geography and Empire* Oxford: Blackwell.

Daniels, P.W. and Lever, W.F. (eds) (1996) *The Global Economy in Transition* London: Longman.

Dasgupta, B. (1998) *Structural Adjustment, Global Trade and the New Political Economy of Development* London: Zed.

Davin, D. (1996) Migration and rural women in China: a look at the gendered impact of large-scale migration' *Journal of International Development* 8, 5: 655–65.

Demeny, P. and McNicoll, G. (1998) *Population and Development* London: Earthscan.

Devas, N. and Rakadi, C. (eds) (1993) *Managing Fast-growing Cities: New Approaches to Urban Planning and Management in the Developing World* London: Longman.

Dicken, P. (1998) *Global Shift* London: Chapman.

Dickenson, J., Gould, B., Clarke, C., Mather, S., Prothero, M., Siddle, D., Smith, C. and Thomas-Hope, E. (1996) *A Geography of the Third World* 2nd edition, London: Routledge.

Dixon, C. and Drakakis-Smith, D. (eds) *Economic and Social Development in Pacific Asia* London: Routledge.

Donnellan, C. (ed.) (1996) *Aids – It's Not Over Yet* Cambridge: Independence.

Dos Santos, T. (1970) The structure of dependency' *American Economic Review* 60: 125–58.

Drakakis-Smith, D. (1987) *The Third World City* London: Methuen.

Drakakis-Smith, D. (1992) *Pacific Asia* London: Routledge.

Dunning, R. (1993) *Multinational Enterprises and the Global Economy* Wokingham: Addison-Wesley.

Dwyer, D.J. (1975) *People and Housing in Third World Cities* London: Longman.

Dyker, D.A. (ed.) (1992) *The European Economy* London: Longman.

Elliott, J.A. (1999) *An Introduction to Sustainable Development: The Developing World* 2nd edition, London: Routledge.

Eronen, J. (1998) A geopolitical approach to China's future as an empire' *Tidjschrift voor Economische en Sociale Geografie* 89, 1: 4–14.

Felix, D. (1998) Is the drive toward free-market globalization stalling? *Latin American Research Review* 33, 3: 191–200.

Fine, B. (1999) The developmental state is dead – long live social capital? *Development and Change* 20, 1. 1–19.

Flora, C.B. (1998) Beyond explanation and integration: new scholarship on women in Latin America' *Latin American Research Review* 33, 2: 245–57.

Foucault, M. (1979) *Discipline and Punish* New York: Pantheon.

Founoutchuigoua, B. (1996) 'Africa confronted by the ravages of neoliberalism' *Africa Development* 21: 5–24.

Frank, A.G. (1967) *Capitalism and Underdevelopment in Latin America* New York: Monthly Review Press.

Frankel, J.A. (1999) Soros's split personality' *Foreign Affairs* 78, 2: 1 24–30.

Freeman, D.B. (1996) Doi Moi policy and the small enterprise boom in Ho Chi Minh City, Vietnam' *Geographical Review* 86, 2: 178–97.

Friedmann, J. (1966) *Regional Development Policy: A Case Study of Venezuela* Cambridge, MA: MIT Press.

Fung, K.I. (1980) Suburban agricultural land use since 1949' in Leung, C.K. and Ginsberg, N. (eds) *China: Urbanisation and National Development* Chicago: University of Chicago Press.

Furtado, C. (1969) *Economic Development in Latin America* Cambridge: Cambridge University Press.

Fyvie, C. and Ager, A. (1999) NGOs and innovation: organizational characteristics and constraints in development assistance work in the Gambia' *World Development* 27, 8: 1383–95.

Garten, J.E. (1999) Lessons for the next financial crisis' *Foreign Affairs* 78, 2: 76–92.

German, T. and Randel, J. (eds) (1993) *The Reality of Aid* London: Actionaid.

Gibb, R. and Michalak, W. (1996) Regionalism in the world economy' *Area* 28, 4: 446–58.

Gibb, R. and Michalak, W. (eds) (1993) *Continental Trading Blocs: The Growth of Regionalism in the World Economy* London: Wiley.

Giddens, A. (1999) *Runaway World: How Globalisation is Reshaping Our Lives* London: Profile.

Gilbert, A.G. and Gugler, J. (1992) *Cities, Poverty and Development: Urbanization in the Third World* 2nd edition, Oxford: Oxford University Press.

Gleave, M.B. (ed.) (1992) *Tropical African Development* London: Longman.

Gould, W.T.S. (1993) *People and Education in the Third World* London: Longman.

Gourevitch, P.A. (ed.) (1989) The Paci fic region: challenges to policy and theory', special edition of *Annals of the American Academy of Political and Social Science* 505 (September).

Grainger, A. (1993) *Controlling Tropical Deforestation* London: Earthscan.

Gugler, J. (ed.) (1996) *The Urban Transformation of the Developing World* Oxford: Oxford University Press.

Gwynne, R.N. (1999) Globalisation, commodity chains and fruit exporting regions in Chile' *Tijdschrift voor Economisch en Sociale Geografie* 90, 2: 211–25.

Hall, T. (1998) *Urban Geography* London: Routledge.

Harrison, D. (1988) *The Sociology of Modernisation and Development* London: Routledge.

Harriss, J. and Harriss, B. (1979) Development studies' *Progress in Human Geography* 3, 4: 577–82.

Hettne, B. (1995) *Development Theory and the Three Worlds* London: Longman.

Hirschman, A.O. (1958) *The Strategy of Economic Development* New Haven, CT: Yale University Press.

Hodder, B.W. (1978) *Africa Today* London: Methuen.

Hodder, R.N.W. (1987) Some dynamics of peri-urban vegetable farming in China' MPhil thesis, University of Hong Kong (unpublished).

Hodder, R.N.W. (1989) The Shanghai municipality, 1978–87' PhD thesis, University of Leeds (unpublished).

Hodder, R.N.W. (1991) Planning for development in Davao City, the Philippines' *Third World Planning Review* 13, 2: 105–28.

Hodder, R.N.W. (1992) *The West Pacific Rim* London: Belhaven.

Hodder, R.N.W. (1995) *Merchant Princes of the East* Chichester: Wiley.

Hutchinson, J. and Smith, A.D. (eds) (1996) *Ethnicity* Oxford: Oxford University Press.

Jameson, F. (1984) Postmodernism, or the cultural logic of late capitalism' *New Left Review* 146: 53–92.

Jenkins, R. (1987) *Transnational Corporations and Uneven Development* London: Methuen.

Johnson, H. (1971) A word to the Third World: a western economist's frank advice' *Encounter* 37.

Johnston, R.J., Taylor, P. and Watts, M.J. (1995) *Geographies of Global Change* Oxford: Blackwell.

Kaletsky, A. (1999) Survival of the richest' *The Times*, 4 February: 20.

King, A. (1994) *Urbanism, Colonialism and the World Economy* London: Routledge.

Knox, J. (1995) World cities and the organisation of global space' in Johnston, R.J. *et al.* (eds) *Geographies of Global Change* Oxford: Blackwell, pp. 233–4.

Knox, P.L. and Taylor, P.J. (eds) (1995) *World Cities in a World-System* Cambridge: Cambridge University Press.

LeClair, M.S. (1997) *Regional Integration and the Global Free Trade* Aldershot: Avebury.

Leeming, F. (1993) *The Changing Geography of China* Oxford: Blackwell.

Leftwich, A. (1993) Governance, democracy and development in the Third World' in Corbridge, S. (ed.) *Development Studies: A Reader* London: Arnold.

Lewis, W.A. (1955) *The Theory of Economic Growth* London: Allen and Unwin.

Leys, C. (1996) *The Rise and Fall of Development Theory* London: Currey.

Lipton, M. (1977) *Why Poor People Stay Poor: Urban Bias In World Development* London: Temple Smith.

Little, I. (1982) *Economic Development: Theory, Policy and International Relations* New York: Basic Books.

Lloyd, P. (1979) *Slums of Hope: Shanty Towns of the Third World* Harmondsworth: Penguin.

Lloyd, P.C. (ed.) (1966) *The New Elites of Tropical Africa* Oxford: Oxford University Press.

Lloyd, P.C., Mabogunje, A.L. and Awe, B. (1967) *The City of Ibadan* Cambridge: Cambridge University Press.

Maasdorp, G. (1996) *Can South and Southern Africa Become Globally Competitive Economies?* London: Macmillan.

Mabogunje, A.L. (1989) *The Development Process: A Spatial Perspective* 2nd edition, London: Unwin Hyman.

McGee, T.G. (1967) *South East Asian City* London: Bell.

McGee, T.G. (1994) The future of urbanisation in developing countries: the case of Indonesia' *Third World Planning Review* 16: iii–xii.

Main, H. and Williams, S.W. (eds) (1994) *Environment and Housing in Third World Cities* Chichester: Wiley.

Mandami, M. (1972) *The Myth of Population Control* New York: Monthly Review Press.

Mangin, W. (1967) Latin American squatter settlements: a problem and a solution' *Latin American Research Review* 2: 65–98.

Mauro, P. (1999) Internal IMF paper quoted in *The Economist*, 16 January: 27.

Meadows, D.H., Meadows, D.L., Randers, J. and Behrens, W.W. (1972) *The Limits to Growth* London: Pan.

Mehmet, O. (1995) *Westernising the Third World* London: Routledge.

Mohan, G. (1996) SAPs and Development in West Africa' *Geography* 81, 4: 364–8.

Momsen, J.H. (1991) *Women and Development in the Third World* London: Routledge.

Monan, J. (1995) *Bangladesh* Oxford: Oxfam.

Murphy, A.B. (1995) Economic regionalization and Paci fic Asia' *Geographical Review* 85, 2: 127–40.

Myint, H. (1964) *The Economics of Developing Countries* London: Hutchinson.

Myrdal, G. (1957) *Economic Theory and Underdeveloped Areas* London: Duckworth.

Narasaiah, M.L. (1996) Uruguay Round: impact on Africa' *Africa Quarterly* 4: 81–5.

Narmann, A. and Simon, D. (1999) *Development in Theory and Practice* London: Longman.

Nel, E., Hill, T. and Binns, T. (1997) Development from below in the New" South Africa: the case of Hertzog, Eastern Cape' *Geographical Journal* 163, 1: 57–64.

Nicholls, W.H. (1971) The Brazilian food supply: problems and prospects' *Economic Development and Cultural Change* 19, 3: 387–8.

Nixson, F. (1996) *Development Economics* London: Heinemann.

Nowak, M. and Swinehart, T. (eds) (1989) *Human Rights in Developing Countries*, 1989 Yearbook, Kehl, Strasbourg: N.P. Engel.

O'Connor, A. (1983) *The African City* London: Hutchinson.

O'Connor, A. (1991) *Poverty in Africa: A Geographical Approach* London: Belhaven.

O'Malley, W.J. (1988) Culture and modernization' in Hughes, H. (ed.) *Achieving Industrialization in East Asia* Cambridge: Cambridge University Press.

Parnwell, M. and Turner, S. (1998) Sustaining the unsustainable city: city and society in Southeast Asia' *Third World Planning Review* 20: 147–64.

Peet, R. (1991) *Global Capitalism* London: Routledge.

Perkins, D.H. (1978) Meeting basic needs in the People's Republic of China' *World Development* 6: 596–610.

Peterson, P.G. (1999) Gray dawn: the global aging crisis' *Foreign Affairs* 78, 1: 42–55.

Pieterse, J.N. (1966) The development of development theory' *Review of International Political Economy* 3: 541–4.

Poon, J.P.H. and Perry, M. (1999) The Asian economic " flu," a geography of crisis' *Professional Geographer* 51, 2: 184–96.

Potter, R.B. (1993) Little England and little geography: re flections on Third World teaching and research' *Area* 25: 291.

Potter, R.B. (1997) Third World urbanisation in a global context' *Geography Review* 10: 2–6.

Potter, R.B. and Lloyd-Evans, S. (1998) *The City in the Developing World* London: Longman.

Potter, R.B., Binns, T., Elliott, J.A. and Smith, D. (eds) (1999) *Geographies of Development* London: Longman.

Potts, D. (1995) Shall we go home?Increasing poverty in African cities and migration processes' *Geographical Journal* 161, 3: 245–64.

Prebisch, R. (1950) *The Economic Development of Latin America* New York: United Nations.

Pred, A. (1977) *City-Systems in Advanced Economies* London: Hutchinson.

Preston, P.W. (1996) *Development Theory: An Introduction* Oxford: Blackwell.

Pugh, C. (ed.) (1996) *Sustainability, the Environment and Urbanization* London: Earthscan.

Redding, S.G. (1990) *The Spirit of Chinese Capitalism* Berlin: Walter de Gruyter.

Reed, D. (ed.) (1996) *Structural Adjustment: The Environment and Sustainable Development* London: Earthscan.

Richards, P. (1985) *Indigenous Agricultural Revolution* London: Hutchinson.

Riedel, J. (1988) Economic development in East Asia: doing what comes naturally' in Hughes, H. (ed.) *Achieving Industrialization in East Asia* Cambridge: Cambridge University Press.

Rigg, J. (1991) *Southeast Asia: Region in Transition* London: Unwin Hyman.

Rigg, J., Allott, A., Harrison, R. and Kratz, U. (1999) Understanding languages of modernization: a Southeast Asian view' *Modern Asian Studies* 33, 3: 581–602.

Rodney, W. (1972) *How Europe Underdeveloped Africa* Dar es Salaam: Tanzanian Publishing House.

Rostow, W.W. (1960) *The Stages of Economic Growth: A Non-communist Manifesto* Cambridge: Cambridge University Press.

Said, E. (1979) *Orientalism* New York: Village Books.

Said, E. (1993) *Culture and Imperialism* London: Chatto.

Sassen, S. (1999) Global financial centers', *Foreign Affairs* 78, 1: 75–87.

Savage, M. and Warde, A. (1993) *Urban Sociology, Capitalism and Modernity* London: MacmillanBritish Sociological Association.

Schroeder, R.A. (1999) Community, forestry and conditionality in the Gambia' *Africa* 69, 1: 1–22.

Schumpeter, J.A. (1934) *The Theory of Economic Development* Cambridge, MA: Cambridge University Press.

Sen, A.K. (1994) Population: delusion and reality' *New York Review of Books* 41, 15: 1–8.

Sharp, R. (1994) *Senegal: A State of Change* Oxford: Oxfam.

Sheahan, J. (1998) Changing social programmes and economic strategies: implications for poverty and inequality' *Latin American Research Review* 33, 2: 185–96.

Shen, J. (1998) China's future population and development challenges' *Geographical Journal* 164, 1: 32–40.

Slater, D. (1993) The geopolitical imagination and the enframing of development theory' *Transactions of the Institute of British Geographers* New Series 18: 419–37.

Smith, D.W. (1998) Urban food systems and the poor in developing countries' *Transactions of the Institute of British Geographers* New Series 23, 2, 207–19.

Soros, G. (1998) *The Crisis of Global Capitalism* London: Little, Brown.

Stallings, B. (1995) *Global Change, Regional Response: The New International Context of Development* New York: Cambridge University Press.

Streeten, P. (1995) *Thinking about Development* Cambridge: Cambridge University Press.

Sundberg, J. (1998) NGO landscapes in the Maya biosphere reserve, Guatemala' *Geographical Review* 88, 3: 388–412.

Suryadinata, L. (1989) *The Ethnic Chinese in ASEAN States* Singapore: ISEAS.

Sweetman, C. (ed.) (1998) *Gender and Migration* Oxford: Oxfam.

Todaro, M.P. (1997) *Economic Development* 6th edition, London: Longman.

Tomlinson, J. (1999) *Globalization and Culture* Oxford: Blackwell.

Tordoff, W. (1992) The impact of ideology or development in the Third World' *Journal of International Development* 4, 1: 41–53.

UNDP (1992) *Human Development Report 1992* Oxford: Oxford University Press.

UNDP (1998) *Human Development Report 1998* Oxford: Oxford University Press.

Von Moltke, K. (1994) 'The World Trade Organisation: its implications for sustainable development' *Journal of Environment and Development* 3, 1: 43–57.

Wade, R. (1990) *Governing the Market: Economic Theory and the Role of Government in East Asian Industrialisation* Princeton, NJ: Princeton University Press.

Walker, L. (1995) 'The Uruguay Round: in whose interests? *Review of African Political Economy* 22: 564–8.

Wallerstein, E. (1979) *The Capitalist World Economy* Cambridge: Cambridge University Press.

Ward, R.M. and Liang, W. (1995) 'Shanghai water supply and wastewater dispersal' *Geographical Review* 85, 2: 141–56.

Warren, B. (1980) *Imperialism: Pioneer of Capitalism* London: Verso.

Webster, A. (1990) *Introduction to the Sociology of Development* London: Macmillan.

Wei, Y.D. (1999) 'Changes in industrial ownership structure in China' *Geography* 84, 3: 193–203.

Williams, S. (1998) *Tourism Geography* London: Routledge.

Winchester, S. (1991) *The Pacific* London: Hutchinson.

World Bank (1992) *World Development Report, 1992* Oxford: Oxford University Press.

World Bank (1997a) *World Development Report, 1997: The State in a Changing World* Oxford: Oxford University Press.

World Bank (1997b) *Selected World Development Indicators* Washington, DC: World Bank.

World Bank (1998a) *Reducing Poverty in India* Washington, DC: World Bank.

World Bank (1998b) *World Bank Atlas, 1998* Washington, DC: World Bank.

World Bank (1999) *World Development Report, 1998/99* Oxford: Oxford University Press.

Wuyts, M., Mackintosh, M. and Hewitt, T. (eds) (1992) *Development Policy and Public Action* Oxford: Oxford University Press.

Yapa, L. (1998) 'The poverty discourse and the poor in Sri Lanka' *Transactions of the Institute of British Geographers* New Series 23, 1: 95–115.

Yeung, H.W.-C. (1998) 'Capital, state and space: contesting the borderless world' *Transactions of the Institute of British Geographers* New Series 23, 3: 291–309.

Young, E.M. (1996) *World Hunger* London: Routledge.

Zhou Shulian and Wang Haibo (1982) various chapters in Lin Wei and Arnold Chao (eds) *China's Economic Reforms* Philadelphia: University of Pennsylvania Press.

Index

Printed and bound by CPI Group (UK) Ltd, Croydon, CR0 4YY

01/11/2024

01782621-0006